PRAISE FOR
Master of Change

"Change is inevitable. Will you let this reality destabilize or empower you? Brad Stulberg's immensely wise and timely book provides a powerful roadmap for a tumultuous world."
—Cal Newport, *New York Times* bestselling author of *Digital Minimalism* and *Deep Work*

"According to the Greek philosopher Heraclitus, you can't step in the same river twice. Brad Stulberg makes you wonder why you'd want to. Change is inevitable, but if we learn to struggle *with* it, not *against* it, we become like rivers ourselves: fluid, flexible, yet ruggedly defined. This book speaks to the twists and eddies in our lives with uncommon warmth and wisdom."
—Kieran Setiya, author of *Life Is Hard: How Philosophy Can Help Us Find Our Way*

"Fantastic, wide-ranging, and deeply profound, Brad Stulberg's new book is a reminder that the most important two-word summary of life is: everything changes. In a world of constant economic, technological, and political upheaval, it's more important than ever that we learn to face life with tragic optimism—to continuously rediscover meaning and strength despite life's inevitable challenges and suffering. Stulberg's book is an exceptional guide to this critical virtue. A must read."
—Derek Thompson, staff writer at *The Atlantic*, host of *Plain English* podcast, and author of the national bestseller *Hit Makers*

"*Master of Change* challenges our inherent resistance to change, providing carefully researched suggestions that will help you embrace and rise to meet it. Stulberg's thoughtful book is a delight to read and will arm you with valuable wisdom for navigating life's unexpected plot twists."

—Katy Milkman, *New York Times* bestselling author of *How to Change*

"Uncertainty and changeability are the reality for us all. This is a fascinating and inspiring field guide not merely to 'coping with' this reality, but plunging into it—even becoming one with it—so as to grow and flourish here and now."

—Oliver Burkeman, author of *Four Thousand Weeks*

"In a world where constant flux is the only certainty, the ability to reframe and adapt to change—rather than resist or deny it—is essential to well-being. In *Master of Change*, Brad Stulberg wields his characteristically deft mix of science and philosophy to provide a blueprint for embracing the unexpected."

—David Epstein, author of *Range*

Master *of* Change

Also by Brad Stulberg

The Practice of Groundedness

The Passion Paradox with Steve Magness

Peak Performance with Steve Magness

Master of Change

The Case for Rugged Flexibility

Brad Stulberg

HarperOne
An Imprint of HarperCollinsPublishers

Without limiting the exclusive rights of any author, contributor or the publisher of this publication, any unauthorized use of this publication to train generative artificial intelligence (AI) technologies is expressly prohibited. HarperCollins also exercise their rights under Article 4(3) of the Digital Single Market Directive 2019/790 and expressly reserve this publication from the text and data mining exception.

To protect their identities, when requested I have changed the names and other identifying details of many of the nonpublic figures in this book. Everything else about their stories is reported as it happened and is fact-checked accordingly.

MASTER OF CHANGE. Copyright © 2023 by Bradley Stulberg. All rights reserved. Printed in the United States of America. No part of this book may be used or reproduced in any manner whatsoever without written permission except in the case of brief quotations embodied in critical articles and reviews. For information, address HarperCollins Publishers, 195 Broadway, New York, NY 10007. In Europe, HarperCollins Publishers, Macken House, 39/40 Mayor Street Upper, Dublin 1, D01 C9W8, Ireland.

HarperCollins books may be purchased for educational, business, or sales promotional use. For information, please email the Special Markets Department at SPsales@harpercollins.com.

harpercollins.com

FIRST HARPERONE PAPERBACK PUBLISHED IN 2025

Designed by Bonni Leon-Berman

Library of Congress Cataloging-in-Publication Data is available upon request.

ISBN 978-0-06-325317-9

25 26 27 28 29 LBC 5 4 3 2 1

For Caitlin

Contents

Introduction: Rugged Flexibility—A New Model for Working with Change and Thinking About Identity over Time ... 1

PART 1: RUGGED AND FLEXIBLE MINDSET ... 19
Chapter 1: Open to the Flow of Life ... 21
Chapter 2: Expect It to Be Hard ... 45

PART 2: RUGGED AND FLEXIBLE IDENTITY ... 71
Chapter 3: Cultivate a Fluid Sense of Self ... 73
Chapter 4: Develop Rugged and Flexible Boundaries ... 100

PART 3: RUGGED AND FLEXIBLE ACTIONS ... 123
Chapter 5: Respond Not React ... 125
Chapter 6: Making Meaning and Moving Forward ... 154

Conclusion: Five Questions and Ten Tools for Embracing Change and Developing Rugged Flexibility ... 184

Acknowledgments ... 199
Appendix ... 202
Additional Suggested Reading ... 203
Notes ... 206
Index ... 220

Introduction: Rugged Flexibility— A New Model for Working with Change and Thinking About Identity over Time

"The ground was shifting beneath my feet," remembers Thomas, a longtime coaching client of mine. "Things felt out of control."

As was the case for many, 2020 through 2022 were particularly challenging years for Thomas, a forty-five-year-old father of two who works at a professional services firm. In the span of a few months, he was forced to work from home; lost his biggest client; essentially started homeschooling his children, whose district went remote; witnessed his wife get laid off from her job; lost an uncle to the coronavirus; and was unable to spend much in-person time with his father, who died of cancer at the beginning of 2022. "So much changed in such a short period of time. It was disorienting and hard to keep up."

Thomas's story was common during the period in which

a novel coronavirus ravaged the globe, leaving significant human and economic destruction in its wake. It disrupted how we work, play, love, grieve, and participate in our communities. The coronavirus pandemic represents what I call a *disorder event*—something that fundamentally shifts our experience of ourselves and the world we inhabit, sometimes for better and sometimes for worse. The pandemic may be the most recent large-scale disorder event, but it is certainly not the first, nor will it be the last.

Communally, not a decade goes by that we don't experience dramatic disruptions. Examples include war, the emergence of technologies like the internet and more recently artificial intelligence, social and political unrest, economic recession, and environmental crises, all of which are intensifying rapidly. Individually, disorder events are even more common. Examples include starting a job, leaving a job, getting married, getting divorced, having children, losing a loved one, becoming ill, moving to a new town, graduating from school, meeting a new best friend, publishing a book, earning a big promotion, becoming an empty nester, retiring, and so on. Research shows that, on average, people experience thirty-six disorder events in the course of their adulthood—or about one every eighteen months. This does not include aging, the ever-present, ongoing disorder event that many of us futilely resist and deny. We tend to think that change and disorder are the exceptions when, in reality, they are the rules. Look closely and you'll see that everything is always changing, including us. Life *is* flux.

In the few years preceding this book, I published *The Practice of Groundedness,* had a second child, left a secure job, moved across the country, stopped participating in a sport that, for years, had been an outsized source of my identity, had surgery on my leg, and became painfully estranged from certain family members. Notice this list of changes includes a mix of the good and the bad: it wasn't just that hard things happened in this short period of time, it's that *a lot* of things happened in this short period of time. It was overwhelming, and it made for an interesting, if not challenging, few years.

Whenever I shared any of these major life experiences with clients, colleagues, friends, and neighbors they immediately empathized with the fact that I found it at least somewhat disorienting. I learned I was not alone: just about everyone experiences the doubt, fear, and bewilderment that come with recognizing, up close and personal, that life is not as stable as we think or would like it to be.

Herein lies a problem. A central narrative in our culture urges us to seek stability, yet this doesn't reflect the reality that change is constant—and that, with the right skills, it can be a dramatic force for growth. It's time to flip that script. Accepting the inevitability of it can be scary at first, but as I've come to see, and as you'll read throughout this book, embracing life's fluidity actually turns out to be empowering and even an advantage. No doubt, change can hurt, but it comes with all sorts of benefits too.

Learning an entirely new way to conceive of and work with change—what I've come to call "rugged flexibility"—minimizes distress, restlessness, and angst while promoting deep happiness and lasting fulfillment. It also leads to better and more sustainable performance on the activities and pursuits about which we care most. In this way, rugged flexibility is foundational to sustainable excellence: *doing good and feeling good in a way that supports our long-term goals*. Equally important is that skillfully working with change makes for kinder and wiser people—something the world desperately needs. Plus, it's not like we have any choice in the matter: though you may wish it weren't so, you simply cannot pause time or control life. Trying to is a fool's errand, utterly exhausting and a common cause of burnout and languishing for otherwise healthy people.

The latest findings from psychology, biology, sociology, philosophy, and cutting-edge neuroscience all demonstrate that change itself is neutral. It becomes negative or positive based on how we view it and, more importantly, what we do with it. Meanwhile, unlike in the modern-day West, where we view life as linear and relatively stable, many of the world's ancient wisdom traditions, such as Buddhism, Stoicism, and Taoism, recognize the cyclical nature of reality and the pervasiveness of change. Ancient wisdom and modern science agree that impermanence is an undeniable reality, a fundamental truth of the universe. Clinging to the illusion of permanence, hoping that we won't be struck by change, that we'll stay more or less the same, is misguided at best and leads to suffering at worst. Life is an ongoing and oscillating series of ebbs and flows. Developing a strong, enduring, and cohesive sense of self demands specific skills to ride the waves. Unfortunately, these skills aren't typically taught in school, and they have been largely neglected by recent generations, who far too often have deluded themselves with ideas of control, security, and constancy—all of which work fine . . . until they don't.

So here we are. My central goal in this endeavor is to look across science, wisdom, history, and practice to put forward a comprehensive framework that encompasses the foundational qualities you need to not just survive, but thrive, amidst change and disorder. I call this Rugged Flexibility. But before we dive into this new way to conceive of and work with change, it is helpful to understand how we wound up where we are today: a place where we fear instability and volatility, and feel underpowered in their midst. Knowing how we got to where we are will help us get to where we want to go.

Why—and How—We Get Change Wrong

In 1865, a fifty-two-year-old French physician named Claude Bernard arrived at a breakthrough insight. Based on his observations of the human body, he proposed a model that viewed change and disruption as antithetical to health. "The fixity of the internal environment is the condition for free life," he explained to his many followers in the then-burgeoning scientific community. It wasn't until more than sixty years later, in 1926, that an American scientist named Walter Cannon officially coined the term *homeostasis*.

Most people are aware of homeostasis, even if they are not scientists. The word's origin comes from the Greek *homoios*, which means "similar" or "same," and *stasis*, which means "standing." Its modern definition is "the tendency of living systems to resist change in order to maintain stable, relatively constant internal environments." Homeostasis describes a cycle of order, disorder, and order. It states that a system has stability at X, an event occurs that causes disorder and moves the system to chaos and uncertainty at Y, and then the system does everything possible to get back to stability at X, and as swiftly as it can. For example, if you become ill, your body may develop a fever, but numerous processes work to return your temperature to its baseline of 98.6 degrees Fahrenheit.

In some narrow instances (like that of a fever) homeostasis is an accurate model; but as you'll soon see, in many others it is not. Even so, homeostasis has been adopted as the predominant way to think about change in nearly all domains. If you search "homeostasis" and "change" on the internet, you'll discover countless articles on topics as diverse as weight loss, writer's block, quitting smoking, starting a new workout program, and transforming your company's culture. All are written in the spirit of "overcoming homeostasis" and "fighting against" a deep and universal resistance to change.

Homeostasis's long history and simple, intuitive appeal have shaped how people, organizations, and even entire cultures think about change. It is responsible for the fact that we generally view external changes as undesirable and the changes we want to initiate in ourselves as going against a preordained order. While in some cases the experience of change as abnormal may be unavoidable, in the vast majority it is not.

Nevertheless, as a result of our longstanding bias, most people tend to respond in one of four ways when faced with change, whether to themselves or to the broader structures in their lives.

1. **Attempt to Avoid Change or Refuse to Acknowledge It**

    We try to insulate ourselves from what is happening around us, sometimes going as far as to deny change altogether. Examples include the company that refuses to shift to a digital business model, the aging basketball player who doubles down on strengths that benefited her during her prime (but no longer do), the man in a broken relationship who refuses to see its problems, or the think tank that cherry-picks its data to avoid facing reality.

2. **Actively Resist Change**

    We try to stop change from happening, doing everything we can to push back even if change is inevitable and its force overwhelming. Examples include the tennis player who keeps putting off surgery though he has degenerating cartilage in both knees, the company that goes to Capitol Hill to lobby against clean-air regulations instead of innovating, the new parent who takes fruitless measures to preserve her usual nine hours of sleep, the older parent who still wants to tell her college-age daughter what to eat and wear, or the forty-five-year-old who obsessively uses all manner of products in hopes of "reversing" aging.

3. **Sacrifice Agency Amidst Chaos**
We perceive change as something that happens *to* us, and thus relinquish all control over the situation. This is the man who receives a scary health diagnosis and immediately stops paying attention to his diet; the woman experiencing anxiety who refuses to get help, telling herself this is just how it will always be; the person who falls victim to the 24–7 news cycle instead of taking control of their attention; the policy makers who throw their hands up at a problem instead of doing something about it; or the company that reacts in a scattershot manner to an increasingly remote workforce rather than developing a thoughtful and deliberate strategy.

4. **Try to Get Back to Where We Were**
We think about what life was like before a disorder event, comparing and contrasting our new situation to the old and defaulting to attitudes and behaviors that served us in the past. This is the man who gets married but still wants to make all his own decisions, the woman who loses her job in a shrinking industry but expects to get rehired into the same role elsewhere, the family that moves across the country and immediately compares every new person they meet to their best friends from back home, or the company that is forced to lay off 20 percent of its workforce but the next day acts as if nothing happened.

Perhaps not all of the above examples resonate, but you can probably recognize some of these tendencies in yourself, your workplace, your family, or your community. There is a strong propensity to over-index on grasping the old order instead of opening to the possibility of something new. While the above strategies may feel good in the moment, they almost always create problems in the long run.

A New Model for Navigating Change and Disorder

In the late 1980s, two researchers—one a neuroscientist, physiologist, and professor of medicine at the University of Pennsylvania, and the other an interdisciplinary scholar with a focus on biology and stress—observed an interesting phenomenon. In the vast majority of situations, healthy systems do not rigidly resist change; rather, they adapt to it, moving forward with grace and grit. This observation is true whether it is an entire species responding to a shift in its habitat, an organization responding to a change in its industry, or a single individual responding to a disorder event in her life or an ongoing process such as aging. Following disorder, living systems crave stability, but they achieve that stability somewhere new. Peter Sterling (the neuroscientist) and Joseph Eyer (the biologist) coined the term *allostasis* to describe this process. Allostasis comes from the Greek *allo*, which means "variable," and *stasis*, which, as you learned earlier, means "standing." Sterling and Eyer defined allostasis as "stability through change."

Whereas homeostasis describes a pattern of order, disorder, order, allostasis describes a pattern of order, disorder, *re*order. Homeostasis states that following a disorder event, healthy systems return to stability where they started: X to Y to X.

Allostasis states that healthy systems return to stability, but somewhere new: X to Y to Z.* Homeostasis is largely a misnomer. Everything is changing always, including us. We are constantly somewhere in the cycle of order, disorder, *re*order. Our stability results from our being able to navigate this cycle, or as Sterling and Eyer put it, "We achieve stability through change." I interpret this phrase to have a double meaning: the way to stay stable through the process of change is *by* changing, at least to some extent.

To drive the concept home, let's move from a bird's-eye view of allostasis to some simple and concrete examples: If you start lifting weights or gardening regularly, the skin on your hands will almost always become disturbed. Instead of futilely trying to stay smooth, eventually it will develop calluses so it can better meet the challenge. If you are accustomed to constantly shifting your attention in a digital world, your brain will, at first, resist reading a book with no distractions. But if you stay at it, eventually your brain adapts and rewires itself for focus, which scientists call *neurogenesis* or *neuroplasticity*. Still another example is experiencing depression or heartbreak. Recovery is not going back to how you were before you experienced intense psychic pain. Rather, it is moving forward, usually with a greater tolerance for emotional distress and increased compassion for others who are suffering. In these examples you achieve stability not by fighting change or getting back to where you were, but rather by skillfully working with change and arriving at someplace new.

---

* Some scientists now use the term *homeostatic upregulation* to describe components of allostasis, but we are going to stick with the latter terminology as to minimize confusion between the old and new models of change.

"The key goal of regulation is not rigid constancy," writes Sterling. "Rather, it is the flexible capacity for adaptive variation."

Once You Are Aware of Allostasis, You Start to See It Everywhere

Sterling and Eyer first described the basic tenets of allostasis in 1988, yet the concept is still little known among laypeople. This is unfortunate, because it turns out that allostasis is the most accurate and beneficial model for representing change and how our identities evolve and grow over time. The following examples show its profound universality.

Evolution, the grand theory of natural science, is the process by which life advances via adapting to continually changing circumstances. There is no going back to the way things were. Change is a constant. Species that adapt thrive and endure. Species that resist suffer and die out.

In literature, the "hero's journey" describes the predominant theme in myths from across cultures and eras. The hero begins in a stable home environment; experiences a major change or disorder event; is forced to leave their stable home environment; ventures out into a new world where they face obstacles and challenges; and eventually returns home, with a sense of self that is the same but also transformed. This archetype describes myths and stories ranging from Moses of the Israelites, to Siddhartha Gautama of Buddhism, to Simba of *The Lion King* and Mirabel of *Encanto*.

One of the founders of modern psychology, Carl Jung, used a circle to represent the ongoing transformation of the self, arguing that the process of individual becoming is one of perpetual adaptation and growth. Since then, newer therapeutic models, such as cognitive behavioral therapy (CBT) and acceptance and commitment therapy (ACT), teach people not to resist impermanence or try to get back to where they were, but rather to open up to impermanence, work with it, and transcend it.

The Franciscan friar Richard Rohr teaches that we become our truest selves through rounds of order, disorder, and reorder. He goes as far as to call this the *universal wisdom pattern*. The Buddhist teacher and psychotherapist Mark Epstein writes that freedom from anxiety requires learning how to navigate inevitable cycles of integration, un-integration, and *re*integration—what he calls *going to pieces without falling apart*.

In organizational science, researchers describe successful change as a pattern of freezing, unfreezing, and *re*freezing. The unfreezing period is often chaotic, but it is a necessary step to arrive at a stable and enhanced end point. Meanwhile, relational therapists talk about cycles of harmony, disharmony, and repair as the key to growth in all of our important bonds.

Happy, healthy, and sustainably performing individuals and organizations also exhibit this pattern. They maintain a strong and enduring identity by repeatedly remaking themselves. They have the courage to abandon their current station, enter into disorder, and arrive at an enhanced stability and sense of self down the road. What they all hold in common is a view of identity as both stable and changing at the same time.

A Few Ways of Representing Ongoing Cycles of Change and Progress

- Order → Disorder → Reorder
- Stability at X → Chaos and uncertainty at Y → Stability at Z
- Integration → Un-integration → Reintegration
- Orientation → Disorientation → Reorientation
- Freezing → Unfreezing → Refreezing
- Harmony → Disharmony → Repair

A guiding tenet in my work, as both a writer and a coach, is pattern recognition. I'm not interested in "hacks," quick fixes, or single small studies, all of which tend to be high on promises but low on real-world efficacy. Regardless of what the marketers, clickbait headlines, and pseudoscience evangelists say, there are no magic lotions, potions, or pills when it comes to genuine excellence, lasting well-being, and enduring strength. What I am interested in is convergence. If multiple fields of scientific inquiry, the world's major wisdom traditions, and the practices of people and organizations that have demonstrated excellence and fulfillment over time all point toward the same truths, then those truths are probably worth paying attention to. In this instance, change and impermanence are not phenomena to fear or resist—at least not as a default position. Though the historical concept of homeostasis has deeply penetrated our collective psyche, it is an outdated model for navigating life, supporting mental health, and pursuing genuine excellence. Allostasis makes a lot more sense.

Rugged Flexibility

When I first confronted the ubiquity of change, it made me uncomfortable. I am a person who craves stability. I like to have a plan and stick to it. If you were to draw a line with stability on one end and change on the other, you could plot me about a millimeter (and that would be generous) away from the stability extreme. Yet, as I continued down my own life's path, experiencing all manner of volatility, and as I started researching for this book, it occurred to me that no such line exists. Here's why this discovery is powerful: In addition to the major vicissitudes of life we touched on earlier—aging, illness, relationships, relocating, social unrest, and so on—once I began to practice the tenets of rugged flexibility that you'll learn about in the coming pages (emphasizing "practice," since this will be a lifelong process for all of us), I started feeling less restless and worried about the smaller stuff, too. I wasn't as fazed by unplanned changes in my work. I became less frustrated and thrown off when my so-called "perfect" schedule was blown up by a sick kid home from school, a dog with diarrhea, an internet outage, and all manner of other proverbial pebbles in my shoes. When I experienced complications after a surgery and my rehabilitation time doubled, it didn't upset me as much as it once would have. Though these things seem—and in most cases, are—relatively trivial, they add up, and they leave many of us feeling chronically frustrated and showing up short of our potential. Just think about how many bad days at work, arguments with your significant other, and sleepless nights, at their root, result from feelings of distress caused by uncertainty and change.

While some things in life truly are either/or—you are either driving the speed limit or you are not; you are either pregnant or you are not—many are both/and. For example, decision-making is not about reason *or* emotion; it is about reason *and* emotion. Toughness is not about self-discipline *or* self-compassion; it is about self-discipline *and* self-compassion. Progress in just about any endeavor is not about hard work *or* rest; it is about hard work *and* rest. Philosophers call this kind of thinking *non-dual*. Non-dual thinking recognizes that the world is complex, that much is nuanced, and that truth is often found in paradox: not this *or* that, but this *and* that. Non-dual thinking is an important, albeit spectacularly underused, concept in many facets of life, including our subject matter here. As such, it will come up repeatedly throughout this book.

When you apply non-dual thinking to stability and change,

an interesting thing happens. The goal is not to be stable and therefore never change. Nor is the goal to sacrifice all sense of stability, passively surrendering yourself to the whims of life. Rather, the goal is to marry these qualities to cultivate what I call *rugged flexibility*. To be rugged is to be tough, determined, and durable. To be flexible is to consciously respond to altered circumstances or conditions, to adapt and bend easily without breaking. Put those together and the result is a gritty endurance, an anti-fragility that not only withstands change, but thrives in its midst. This is rugged flexibility, the quality you need to become a master of change, to successfully navigate disorder and chaos and endure over the long haul.

Rugged flexibility recognizes that after disorder there is no going back to the way things were—no more order, only *re*order. The goal of rugged flexibility is to get to a favorable reorder; to maintain a strong core identity, but at the same time, to adapt, evolve, and grow. Unlike old ways of approaching change, rugged flexibility conceives of change not as an acute event that happens to you, but rather as a constant of life, a cycle in which you are an ongoing participant. Via this transformative shift, you come to view change and disorder as something you are in conversation with, an ongoing dance between you and your environment. The more skilled you become at this dance, the happier, healthier, and stronger you will be.

The Principles of Rugged Flexibility

During the preceding years, I've spent countless hours thinking about how rugged flexibility can help us master change and how best to develop it. I've read thousands upon thousands of pages of philosophy and psychology, scoured the latest neuroscience research, and interviewed hundreds of experts from across disciplines. I've gone on this journey for myself, for my coaching clients, and for you. The rest of this book lays out what I've found. It is divided into three parts: "Rugged and Flexible Mindset," "Rugged and Flexible Identity," and "Rugged and Flexible Actions." In each part, I will detail the essential, evidence-based qualities, habits, and practices upon which rugged flexibility is founded.

In part 1, we'll learn how to develop a *rugged and flexible mindset*. This will help us cultivate a more harmonious relationship with change. We'll explore why change is often discombobulating, the difference between *having* and *being*, and why the insight of impermanence may be scary at first but is ultimately empowering. We'll dive into fascinating new research on the workings of consciousness and learn how to nurture a crucial sensibility called tragic optimism. We'll see how a rugged and flexible mindset requires opening up to the flow of life and expecting it to be hard, which, paradoxically, makes change, which is to say *life*, just a bit easier.

In part 2, we'll learn how to develop a *rugged and flexible identity*. This will help us to make sense of ourselves when we, and everything around us, are always changing. We'll investigate age-old teachings on selfhood; explore both why the ego is so defiant toward change and how the ego's strength can actually be helpful (that is, until it gets in the way); and dig into the latest research on complexity theory, systems thinking, and ecology. We'll see that diverse fields of inquiry converge around two important themes: developing a fluid sense of self and cultivating rugged and flexible boundaries for your unfolding path.

In part 3, we'll learn how to take *rugged and flexible actions*. While being in conversation with change means relinquishing some agency, it does not mean relinquishing all of it. We cannot control what happens to us, but we can control what we do as a result. We'll learn about fascinating new research showing that the core of what makes us who we are lies not in our thoughts but in our feelings and the behaviors that give rise to them. We'll dive deep into the neuroscience of character, learning that while the brain's hardware is fairly rigid, its software is highly malleable, updating based on the actions we take, particularly in emotionally charged situations. This is welcome news: it allows us to respond rather than react and to turn struggle into meaning.

A Road Versus a Path

Before we dive into the heart of this book, let's briefly consider the difference between a road and a path, which will serve as a useful metaphor going forward. A road is linear and aims to get you from here to there with as much haste and as little effort as possible. A road resists the landscape; instead of working with its environment it plows over whatever is in its way. When you are traveling on a road, you know your destination. If you get knocked off, it is an unambiguously bad thing; you get back on and assume smooth travel again. Interesting opportunities may be calling you from the sides, but when you are on a road, the goal is to stay on the road, to get where you are going as fast as you can.

A path, on the other hand, is quite different. It works in harmony with its surroundings. When you are traveling on a path you may have a general sense of where you are going, but you are open to navigating, perhaps even making use of, whatever detours arise. A path is not separate from its environment but rather part of it. If you get knocked off a road, it can be traumatizing and disorienting. But there is no getting knocked off a path, since it is always unfolding and revealing itself to you. A road resists time and the elements, building up tension until eventually it cracks and crumbles. A path embraces change and is constantly rerouting itself accordingly. Though at first a road may seem stronger, a path is far more robust, durable, and persistent.

Cultivating a strong and enduring sense of self means treating your life like a path. It requires that you do not become too attached to any period of "order" or to any specific route, which usually causes more harm than good and leads to all manner of missed opportunities. Overwhelming science demonstrates that the more distress—what researchers call *allostatic load*—a person, organization, or culture experiences during periods of disorder, the greater their chance of disease and demise. Fortunately, the same science agrees that we can also become stronger and grow from change, and that much about how we navigate it is behavioral; that is, it can be developed and practiced, which is what the rest of this book is about.*

---

* I first considered the difference between a road and a path after reading Wendell Berry's 1968 essay "A Native Hill," which examines the difference between the literal landscapes of roads and paths.

Part 1

RUGGED AND FLEXIBLE MINDSET

1

Open to the Flow of Life

It was shaping up to be the trip of a lifetime, but for reasons no one could have imagined. Tommy Caldwell, a professional climber in his early twenties, was exploring the remote mountains of Kyrgyzstan with his partner Beth Rodden, close friend Jason Smith, and photographer John Dickey, all of whom were also skilled climbers. They were a few days into their adventure and deep in the Kara Su Valley, a part of Kyrgyzstan with towering vertical rock faces reminiscent of Yosemite National Park. After a long stretch of arduous climbing and in desperate need of reprieve, they decided to bivouac on a portaledge. Bivouacking means camping without cover, and a portaledge is an aluminum platform that hangs off the side of a mountain. Suspended under the stars, side by side with the love of his life, Caldwell was elevated physically, emotionally, and spiritually. The group began to decompress and relax, looking forward to a restful evening. But as the old Yiddish adage goes, *Mann tracht, un Gott lacht*: man plans, and God laughs.

It wasn't long after they had settled in that the climbers heard a series of gunshots coming from below. At first they assumed it was a skirmish among rebels, something not uncommon in the region. Their thinking quickly changed as bullets began ricocheting off the rocks around them. Their peaceful portaledge had become a target.

Caldwell, Rodden, Dickey, and Smith decided their best hope was to negotiate with the gunmen. They chose to send down Dickey, the oldest of the group at a mere twenty-five. As soon as he finished repelling to the base of the mountain, Dickey was confronted by three armed members of the Islamic Movement of Uzbekistan, a militia fighting for the creation of an independent Muslim state in Kyrgyzstan.

The rebels spoke hardly any English, but they made clear there was no room for negotiation about letting the Americans go. Dickey summoned the remaining trio of climbers down the mountain, and so began their experience as captives. If there was any uncertainty as to the precariousness of their situation, it ended a few hours into their internment, when the rebels shot and killed a Kyrgyz soldier whom they had also taken hostage.

For the next five days the group traversed the mountainous terrain at gunpoint, walking throughout the night, hiding during the day, and moving in and out of enclaves where they would not be spotted by members of the Kyrgyz military. Caldwell and his friends were offered no food and forced to drink contaminated water. They were cold, sick, and starving—hanging on to hope by a thread that, with each passing hour, seemed increasingly likely to break.

By the sixth day, the situation was becoming untenable for everyone, including the rebels, who decided to split up in order to search for food. One rebel was to take the climbers to a secluded part of the mountain, where they would wait out of sight from Kyrgyz military or other potential rescuers. As they began ascending, Caldwell, Rodden, Dickey, and Smith noticed their captor was ill at ease on the rockface. Yes, he was armed, but the higher they climbed, the shakier and more nervous he became. This, Caldwell thought, presented a window of opportunity.

After agonizing over the idea for what seemed like an eternity, Caldwell knew what had to be done—what *he* had to do. The group landed upon a small and jagged cliff with an overhang barely wide enough to hold them. While topography like this was familiar to Caldwell, Rodden, Dickey, and Smith, it was anything but that to the rebel, who appeared more concerned about his footing and exposure than he was about his captives. Caldwell mustered all his strength, physical and psychological, sprung, and pushed the rebel. *THUD. CLACK. THUD.* The rebel fell, hit a ledge, tumbled, and disappeared over the cliff into sheer and infinite darkness. Looking back on the experience, Caldwell couldn't believe it. "I had just killed somebody. The whole world came crashing down on me all at once," he remembers.

In the moment, however, there was no time for the climbers to process what had happened. They knew the second captor was lurking nearby, and he would kill them if they didn't run away. The climbers quickly gathered themselves and ran for more than four hours until they finally stumbled upon a Kyrgyz military base. There, they were given food and water, and shortly after, helicopter-lifted out of the mountains and eventually sent back to America.

It wasn't long after Caldwell arrived home in Loveland, Colorado, that the enormity of what had happened began to sink in. "I thought I was an evil person," remembers Caldwell. "I said to Beth, 'How can you love me after I did something like this?'"

Caldwell could not escape the fact that this experience had significantly changed him. He was no longer just a kind, lighthearted, and optimistic climber. He was also someone who had killed another human being. He struggled to integrate this massive disorder event into his narrative, becoming a shell of his former upbeat and energetic self. His friends and family could hardly recognize him. He tried to get on with his life like he had before, minimizing and resisting the immensity of his experience; yet he only became more anxious, distant from others, and disassociated from himself—or at least the self he thought he was, the self who he had been. Though what had happened in the mountains was unthinkably hard, in many ways, the aftermath was harder. Caldwell's identity had changed on a dime. It was a painful and discombobulating experience.

Change Is Discombobulating

Before we delve into how to better handle change, we need to first acknowledge that change is rarely, if ever, easy. Caldwell's story is extreme, but its primary teaching is universal. For many people, change tends to be a source of tumult, overwhelm, and distress, the effects of which are detrimental to our health, relationships, and ability to flourish. Yet as you'll soon see, it is not so much change itself that causes harm, but our slow uptake of it, or in some cases, our downright resistance and refusal. Often, this is our first roadblock. If we are to productively work with change, we've got to see it for what it is—which requires overcoming our conditioned reactions and adopting a mindset that accepts, if not embraces, change as an inevitability. During the remainder of this chapter, I'll argue for this mindset and begin showing you how to develop it. We'll start by examining a groundbreaking study that deployed nothing more than a stopwatch and a deck of cards.

In the mid-twentieth century, Harvard psychologists Jerome Bruner and Leo Postman were interested in how people perceive and react to unexpected changes and incongruities. In what turned out to be a landmark experiment that was published in 1949 in *The Journal of Personality,* Bruner and Postman presented participants with decks of cards that contained anomalies: for example, a red six of spades or a black ten of hearts. One by one, the cards were flashed in front of the participants. While some subjects quickly recognized and described the anomalous cards, others were dumbfounded. If it took the subjects twenty milliseconds to recognize and describe a normal card, it might have taken them one hundred to two hundred milliseconds to recognize and describe the anomalous ones. For those most unwilling to adjust their preconceptions about what playing cards might look like, it took fifteen times the average exposure of a normal card. "I cannot make the suit out. Whatever it is. It didn't even look like a card that time. I don't know what color it is now or whether it is a spade or heart. I'm not even sure what a spade looks like. My God!!" exclaimed one such participant.

In another iconic study, conducted around the same time at the Hanover Institute in Hanover, Indiana, researchers designed goggles with specialized inverted lenses. When subjects put on the goggles, their worlds flipped upside down. As a result, they completely lost their perceptual function and experienced extreme disassociation, disorientation, and even personal crises. Perhaps more than anything, participants in the experiment reported feeling lost.

These two experiments are considered foundational to the field of social science. In the time since they were published, numerous others show that people struggle with unexpected changes, especially when those changes are closely related to one's sense of self. This is true in the safe and controlled setting of a laboratory, and even truer in real life. Think back to Tommy

Caldwell. What was his experience in Kyrgyzstan if not being dealt an unimaginable hand, a life suddenly inverted on its axis?

The world's major philosophical traditions recognize the challenge of change. For more than two and a half millennia, the core aim of Buddhism has been to address the suffering caused by clinging, by holding on too tightly to one's possessions, plans, and self-concept in a world where everything is always in flux. The Sanskrit word *viparinama-dukkha* is translated roughly as "the dissatisfaction that results from clinging amidst change." The entirety of Buddhist philosophy is about lessening this dissatisfaction by learning to accept and work with impermanence.

Around the same time the historical Buddha was developing his teachings, the ancient Chinese philosopher Lao Tzu was penning the *Tao Te Ching*, which would become the bedrock of philosophical Taoism. In it, Tzu depicts life as a dynamic path full of uncertainty and instability whose source is an energetic flow. "If you don't realize the source, you stumble in confusion and sorrow," he writes.

Westward and a few centuries later, foreshadowing what would become the popular Christian serenity prayer, the Stoic philosopher Epictetus introduced his dichotomy of control: While there are some things in life that you can control, there is much that you cannot. Suffering, Epictetus taught, arises from trying to manipulate the latter. More recently, the existentialist philosophers—some of the greatest thinkers of the nineteenth and twentieth centuries such as Jean-Paul Sartre, Søren Kierkegaard, Albert Camus, Friedrich Nietzsche, and Simone de Beauvoir—often spoke of what they called the *existential dilemma*, or the undercurrent of confusion and fear that accompanies living in a vast world where everything is impermanent.

Fast-forward to today and we now know that resisting change results in not only psychological suffering, but physical suffering too. Science shows that when you chronically fight change, your body releases the stress hormone cortisol, which is associated with metabolic syndrome, insomnia, inflammation, muscle wasting, and countless other ailments. Perhaps the one thing that hasn't changed in the past 2,500 years is just how hard change is—and how futile and unhealthy, for mind and body alike, resisting it can be.

Fortunately, the effects of change need not be deleterious, if only we know how to handle it. In the Bruner and Postman experiment, once subjects realized and accepted the anomalous cards as part of a new normal, their distress quickly evaporated. In the Hanover inverted goggles study, if subjects could just get through their initial discombobulation and move from rigidity and resistance to a more relaxed and open state, they began to make sense of their inverted vision and found their bearings once again. Buddhism, Taoism, Stoicism, and existentialism all teach that a good, deep, and meaningful life is possible, even probable, if we can learn to accept and work with the inevitability of unrelenting change. Meanwhile, the same modern science demonstrating the harmful health effects of resisting change also shows that if we can let go of our stubbornness and defiance then change actually promotes health, longevity, and growth. Put all this together and a common theme emerges: the goal is to open to the flow of life and accept change, to assimilate the anomalous cards in our own respective experiences and get comfortable in a world that, at times, may seem upside down.

This isn't easy. If it were, everyone would be doing it. Yet, I suspect most of us know when we are resisting the changes in our lives, whether these are good or bad. If we ask ourselves some version of, *What is really happening right now and what can I do about it?* we tend to know deep down whether we are deluding ourselves or not. Working extra hard to come up with a story or rationale for what is happening instead of answering plainly is a good clue that there is likely some pent-up resistance. Hard as it may be at first, if we can bring ourselves to answer plainly, we'll lift an enormous weight off our shoulders: that of resistance, denial, and delusion. With that weight lifted, we can enter into conversation *with* change instead of having it be something that is *happening to* us. This shift is empowering. It makes us more active participants in our own lives, and it allows us to shape our stories.

A Brief—and Important—Tangent on Progress, Resistance, and Societal Change

At the turn of the fifteenth century, Europeans understood Earth as being at the center of the universe, a belief that was heavily enmeshed in the religious dogma of the time. The Church insisted that God had made man and placed him in the middle. But in the mind of an intrepid mathematician and astronomer, that simply couldn't be the case. God may have been almighty, but the math didn't work.

In 1514, Nicolaus Copernicus shared his elegant model of the universe with a select group of friends. The rising and setting of the sun, the movement of the stars, and the changing of the seasons were not due to heavenly forces. They were due to the fact that Earth was rotating around the sun. It took him another twenty-nine years to finish a final draft of his manuscript, *De Revolutionibus Orbium,* translated to modern English as *On the Revolutions of the Heavenly Spheres*. Sensing his masterwork might cause a backlash, Copernicus decided, purely as a matter of diplomacy, to dedicate the book to the standing Pope, Paul the

Third. To his relief, the Church did not ban it, at least not immediately. Copernicus would not live to see whether his theory was widely accepted. He died in May 1543, two months after it was published.

On the Revolutions of the Heavenly Spheres stayed in circulation long enough for another young astronomer to build upon it. Born in Pisa, Italy, in 1564, Galileo Galilei was fascinated by the heavens from a young age. Across all the materials he read, the theory that made most sense to Galileo was that of Copernicus. By then it was called *heliocentrism*, which translates to "sun at the center." For years, Galileo improved upon and publicized heliocentrism. In 1616, at the height of his intellectual prime, he was issued an injunction by the Church to stop teaching it, lest he face dire consequences.

But Galileo would not be deterred. In 1632 he published *Dialogue Concerning the Two Chief World Systems: Ptolemaic and Copernican,* in which he argued unequivocally for heliocentrism. The book was swiftly banned and Galileo was summoned to the Inquisition, a powerful tribunal set up within the Catholic Church to root out and punish heresy throughout Europe and the Americas. He was sentenced to house arrest, the condition under which he lived for ten years until his death in 1642.

Dialogue remained on the index of prohibited books for 111 years. A heavily censored version was finally released in 1744. The original version was not rereleased until 1835, more than two hundred years after its initial publication. In that same year, the Church's prohibition of *On the Revolutions of the Heavenly Spheres* was finally lifted as well. One can only hope that if heaven does exist, Copernicus and Galileo shared a smile while they watched from above as millions of people read their work, all of whom were on a planet that was, in fact, orbiting the sun.

Thankfully, much *has* changed in the past four hundred

years. The scientific method, which, at its core, is about testing one's preconceptions against reality and being open to change, is now a dominant way of thinking in many parts of the world. Even so, the introduction of novel concepts still brings plenty of disruption and dispute. In his popular book *The Structure of Scientific Revolutions,* philosopher Thomas Kuhn observes that scientific progress follows a predictable cycle: The first stage is normalcy, in which there is general agreement about the way things are. Then someone makes a discovery that upends the established way of thinking, which often leads to a crisis. A period of confusion and unrest ensues—the societal equivalent of wearing the Hanover inverted goggles for the first time—until finally a new paradigm is reached. Essentially, Kuhn depicts scientific progress as a cycle of order, disorder, and reorder.

Think about how fast the process Kuhn describes has occurred, and is still occurring, in regard to COVID. Less than two years after a novel virus spread across the globe, science had untangled the mechanisms of transmission and the virus's DNA, leading to effective vaccines and therapeutics. Could science have worked better and faster? Absolutely. But when you zoom out, compared to where we were a few centuries ago, the acceptance of and response to COVID looks like a miracle. Nonetheless, Kuhn observed that there is almost always a segment of people who resist change until the bitter end, and in doing so, leave plenty of suffering in their wake. Unfortunately, this is something that *hasn't* changed.

Demagogues, authoritarians, and grifters thrive during periods of disorder. They offer a false sense of status and security to those who dislike or feel threatened by what is happening. They represent the past, fighting to go back to the way things were instead of moving forward toward something better. Though this book is not meant to be political, I would be remiss not to mention the resurgence across the globe of strongman leaders who stoke and prey upon ambiguous fears in the populace, including in my own country, the United States of America. In 2016, Americans saw the rise of Donald Trump and Trumpism, a political movement defined vaguely by "making America great *again*" (italics mine).

Though Trumpism may be disturbing, it is not surprising. As we saw earlier in this chapter, many people panic when they are dealt anomalous cards. The rise of LGBTQ and women's rights, the reality of climate change and the subsequent trade-offs we are being called upon to face, reckoning with the legacy of slavery and working toward genuine racial justice, waking up to the gruesome and embarrassing costs of too-lax gun laws, and adjusting to an economy with more technology and automation all represent cards that many Americans could never have imagined. Trumpism and other similar movements capitalize on people's discombobulation, offering them the false hope that by taking part in the movement, they will successfully evade change, maintain their status, and remain strong as they are. Of course, some change *is* worth resisting—the rise of evil forces such as Nazism comes to mind—but fighting against basic science, basic rights, basic decency, and basic liberalism does not make sense, especially not in a society that was founded upon these ideals and that functions well enough because of them.

Is it anything but entirely predictable that in addition to downplaying, resisting, or denying all of the above societal developments, Trumpism also downplayed, resisted, and denied COVID? It's all part of the same underlying syndrome: a rampant fear of change and a total unwillingness to accept it, let alone work with it productively. It is the opposite of rugged and flexible. It is weak and rigid. (For what it's worth, the right doesn't have a monopoly on illiberalism. Certain segments of the political left increasingly shunt open discourse and twist facts, though in my opinion not to the same extent.)

Trumpism and similar movements may make some people feel safer in the short term, but they are a recipe for disaster in the long term, leading to a highly fractured society with significant regressive pockets. Remember, life *is* change. If you fear change, then, in many ways, you fear life—and chronic fear becomes toxic both in self and in the culture at large. If, however, more people had the skills to confront uncertainty and impermanence, we would not have to worry so much about grifters, demagogues, and authoritarian leaders. At the end of the day, we all share the ultimate source of uncertainty and impermanence: our own mortality. If we could face this and lesser contingencies more courageously, if we didn't need scapegoats and strongmen to numb our fears, if instead we could more gracefully accept change, I imagine it would give rise to abundant compassion, belonging, and hope rather than extremism, loneliness, and despair.

Arriving at Acceptance

Tommy Caldwell, who we'll continue to use as a case study in this particular chapter, forged ahead with his life. Though he would never be the same person as he was before Kyrgyzstan, he realized there was still joy to be had and large parts of his story yet to be written. Perhaps more than anything, though, it was scaling big walls that helped Caldwell regain his footing on the ground. Pursuits that align with your core values and shrink a big, unwieldy, and overwhelming world to make it feel smaller and more manageable are useful for integrating significant changes into your life and walking confidently into the unknown—a topic we'll explore more in parts 2 and 3 of this book. For Caldwell, climbing served this purpose. It wasn't a complete escape, for that would be unhealthy—and in this case, impossible. Flashbacks, sporadic feelings of dread, and questions about who he was and what he was capable of still haunted him. But climbing was a part of Caldwell before, during, and after his captivity, a thread that ran through his entire life and thus provided continuity. Not to mention, when you are hundreds of feet in the air solving challenging geometry and physics problems, you've got no choice but to focus on what is in front of you, to be present with what *is*, not with what was or could be.

And so Caldwell climbed, gradually feeling more like himself, though a new version. "My way of dealing with it was you just get back on that horse and go climb again," he says. "I didn't know what to think about Kyrgyzstan. Part of me was feeling empowered. I was like, when shit really hit the fan, I was able to do what needed to be done to get us out of there."

In November 2001, about eighteen months after the Kyrgyzstan trip, Caldwell and Rodden, who had by then moved in with each other, were remodeling their Rocky Mountain home in Estes Park, Colorado. Caldwell, twenty-three years old at the time, was building a platform for the couple's new washer and dryer. He was shaping two-by-fours with a table saw, feeding wood into it lengthwise. Suddenly, a small chunk of debris shot out. As Caldwell went to turn off the saw and investigate what had happened, he noticed a few drops of liquid on the table's black surface. In his book *The Push*, Caldwell recalls, "I raised my left hand. Blood burbled up from the stump of a finger like water from a leaky drinking fountain. I saw the white bony stub of my index finger . . . Panic flooded my mind: How can I climb without a left index finger?"

A wave of dizziness washed over Caldwell. He blinked a few times and took a deep breath. He had to find the finger. "I scanned the saw table, scrambled around the side, careful to keep my hand above my heart while searching the ground. Not wanting to distress Beth, I turned toward the house and called to her, voice steady, *I just chopped off my finger. Please come out here."*

Rodden rushed outside and saw Caldwell's finger lying beside the saw. She snatched it out of a pile of sawdust and tossed it in a ziplock bag full of cold water. They rushed to the nearest hospital in Estes Park. There, doctors injected Caldwell with the numbing agent Novocain, packed the finger on ice, and sent him to the larger and more sophisticated hospital in the neighboring town of Fort Collins, about an hour's drive east. Over the next two weeks, doctors did everything they could to reattach his finger, trying three different surgeries. Ultimately, modern medicine was no match for the table saw and the anatomy of the hand. The complexity of ligaments and nerve endings rendered a successful reattachment all but impossible.

The index finger is essential to rock climbing. The way you grip a rock is by placing your index finger on tiny ledges, what climbers call "holds," and wrapping your thumb over your index finger for additional support. Climbing without an index finger is akin to playing basketball without a hand. It is possible, but hard to imagine doing at an elite level, especially when the loss is sudden and occurs during the prime of one's career, precluding any formative adaptations. Caldwell's doctors told him that he'd have to find a new profession. "Everybody around me except for Beth and my parents looked at me and they're like, 'He's done. That's so sad,'" remembers Caldwell.

Once again, Caldwell's path had changed in an instant. This time around, however, he more quickly identified the anomalous cards he'd been dealt and wasted little time resisting or despairing over them. "When I started climbing again, I felt a surprising amount of exhilaration. My focus and direction were crystal clear," he said. "I realized that it wouldn't help to dwell on what went wrong. I told myself that pain is growth. That the trauma would enhance my focus. I assumed that no one outside of my family truly expected me to make a full comeback, an idea I found strangely liberating." No doubt, he still had plenty of physical and psychological pain, but he accepted that he would no longer have a left index finger, and he got back to work—playing the same game he'd always played, but with his new cards, embodying both ruggedness and flexibility. The climbing was arduous and filled with disappointments and setbacks. Simple moves that Caldwell had once been able to execute in his sleep became complicated and difficult. But he persisted undeterred.

He wasn't sure how far he'd go, and he was under no delusion that his disability wouldn't be an issue. But he accepted his fate and continued on his path. He may have been uncertain of where it would take him, but even that was starting to feel okay. He was dropping the weight of resistance and opening to the flow of life.

Having Versus Being

"If I am what I have and what I have is lost, then who am I?" wrote the polymath Erich Fromm in his penultimate book, *To Have or To Be?*, published in 1976. That the topic was on Fromm's mind is not surprising. He was in his mid-seventies while working on the manuscript. By then, he had been forced out of his homeland and watched the Nazis destroy it; married and divorced; transcended intellectual disciplines and labels of psychologist, psychoanalyst, sociologist, and philosopher; published over twenty books; and, in his later years, suffered multiple serious ailments and bore witness to the decline and death of numerous colleagues and friends. In short, Fromm had lived a full and textured life, which means he had experienced much impermanence and change.

The main argument in *To Have or To Be?* is simple yet profound. When you operate in having mode, you define yourself by what you have. This makes you fragile because those objects and attributes can be taken away at any given time. "Because I *can* lose what I have, I am necessarily constantly worried that I *shall* lose what I have.... I am afraid of love, of freedom, of growth, of change, and of the unknown," writes Fromm. When you operate in being mode, however, you identify with a deeper part of yourself: your essence and core values, your ability to respond to circumstances, whatever they may be. A having orientation is static and intolerant to change. A being orientation is dynamic and open to change. Given the reality of unrelenting change, it is easy to see why the latter is advantageous. Tommy Caldwell's arriving at acceptance required that he go from *having* a plan, *having* a youthful innocence, and *having* an index finger to *being* in conversation with life and working with whatever it threw at him.

Perhaps a more relatable example of the benefits of adopting a *being* orientation is that of my coaching client Christine. In the years preceding the onset of COVID, Christine worked as the marketing director of a fast-growing fitness company that her husband had cofounded with two other people. Her responsibilities were broad and stimulating, ranging from website design, to copywriting, to event planning, to new employee onboarding, to member communication. "The work was hard and the hours were long, but it was the best job I'd ever had," she says.

In March 2020, as the reality of the pandemic was sinking in, the gym was forced to close its brick-and-mortar location. The leadership team scrambled, implementing COVID response plan after COVID response plan, doing everything they could to continue offering value to their clients without an operable physical space. But the longer the shutdown ran, the clearer it became that quarantine wasn't going to be a three-week blip; it was going to be a new, monthslong reality. Overnight, the company had shifted from growth mode to survival mode, which meant it could not afford to keep Christine on payroll. Meanwhile, Christine, who had just taken out a mortgage together with her husband on their first home, couldn't afford to work for free.

"Stepping down from my leadership position felt wrong—like abandonment," Christine told me. "I couldn't bring myself to train at the gym because training just reminded me that I wasn't marketing director of the company anymore. This, of course, made me feel even more unsettled. For almost a decade, lifting weights had been one of my primary ways of dealing with challenges. Now that was off the table too. I was unsure of who I was and my role in the community. I was lost."

For better or worse, she didn't have much time to stew in anguish. She and her husband needed money. It was only a few days into her job search when a voice inside her head asked, *Why don't you try becoming a writer?*

Christine had loved writing for as long as she could remember. Whereas other young kids wanted to be astronauts or doctors or veterinarians, she had always wanted to be an author. Though she'd studied English in college, she told herself when she graduated that writing wasn't a realistic career path. She assumed that in order to earn a living she had to pursue something more reliable. So Christine took jobs that dressed writing up in what she considered to be more practical costumes. First, she became an English teacher and taught composition. Then she became a curriculum designer and wrote lesson plans. Then she became a marketing director and wrote copy and strategic plans. As she sat there during quarantine, her "reliable" and "stable" job having just been unceremoniously ripped away, she realized that while many things in life come and go, her love of writing and her desire to become an author were ever-present forces.

When Christine shared this with me, I asked her what she had to lose. I'd seen her writing and it was solid. I told her that too many people never take shots because they strive for perfection when good enough is usually, well, good enough. It didn't take much encouragement for Christine to open her own copywriting and ghost-writing business. She started chipping away, marketing her work locally and building up a boutique creative writing service.

Today, Christine's writing shop is going strong and she's gained the confidence to start taking big swings at her own creative projects. She's also back at the gym lifting weights. "The fact that I'm actively pursuing my childhood dream is incredible," she explains. "It's challenging in its own right, but I wouldn't give that challenge up for anything, even the marketing job I'd loved so much. Leaving our fitness company was painful, yes, but if I hadn't done that, I wouldn't be doing this. And doing this, to me, is living my life to the fullest."

Christine's *being* orientation was integral to her surviving, and even thriving, amidst so much change and uncertainty. She quickly realized that her identity was more than what she had (her role as marketing director at her company), which enabled her to move forward with ruggedness and flexibility. Central to Christine's being was creativity and her love of the written word, neither of which could be taken away. Toward the end of *To Have or To Be?*, Fromm writes that true joy is "what we experience in the process of growing nearer to the goal of becoming ourselves." Which is precisely what Christine is doing.

The Inescapability Trigger

In addition to letting go of a having orientation in favor of being in conversation with their lives, Caldwell and Christine both benefited from what Harvard behavioral scientist Daniel Gilbert calls the *inescapabilty trigger*. "We are more likely to look for and find a positive view of things we are stuck with than things we are not," writes Gilbert. "Only when we cannot change the experience, and we fully realize that, can we start to change our relationship to the experience." Caldwell knew his finger was not coming back. Christine knew that COVID was not going to disappear overnight. It was clear to both of them that there was no escaping their respective situations.

Once you accept something as an immutable reality in the present moment, you give yourself permission to stop wishing it away or trying to manipulate it on your terms. This allows you to direct all of your energy toward acceptance and moving forward. The key is that you've got to *truly* accept your reality: not think about accepting it, not talk about accepting it, not wish you could accept it, but actually accept it. Your mind all too easily detects when you are lying to yourself. This explains why people in rough situations—for instance, a soul-sucking job—often have to hit a bottom of sorts before they can move forward in earnest. A person can tell themselves that what they are doing is not good for them and that they need to stop, but until they mean it in every bone of their body, their mental and emotional energy goes toward finding a solution in the current state instead of imagining a new state altogether. If, however, we can acknowledge hard truths instead of deluding ourselves, the reward is a significant one: a more meaningful life.

A Deep and Meaningful Life Amidst Impermanence

In 1915, at the beginning of World War I and three years before the 1918 influenza pandemic, Sigmund Freud wrote a short and powerful essay titled "On Transience." While many of Freud's ideas have since been disproven, this essay stands the test of time. It begins with Freud on a countryside walk accompanied by two friends, including a "young but already famous poet," whom many suspect was Rainer Maria Rilke.

"The poet admired the beauty of the scene around us but felt no joy in it. He was disturbed by the thought that all this beauty was fated to extinction, that it would vanish when winter came, like all human beauty and all the beauty and splendor that men have created or may create. All that he would otherwise have loved and admired seemed to him to be shorn of its worth by the transience which was its doom," writes Freud. He went on to explain that while he did not dispute the transience of all things, even the most beautiful and perfect, he did dispute the poet's pessimism and disdain. If anything, Freud argued the opposite: the fact that everything in this world is transient *increases* its value. "Transience value is scarcity value in time. Limitation in the possibility of an enjoyment raises the value of the enjoyment. A flower that blossoms only for a single night does not seem to us on that account less lovely," he wrote. Freud shared all of this with the poet, but it made no difference. The poet couldn't fully experience the beauty around him because that would also require accepting the inevitability of its loss.

The poet's dilemma is common and as old as time. In ancient Sanskrit texts there are two kinds of change, *anatta* and *anicca*. Anatta explains that what you identify as *you* is always changing. Anicca describes the rapidly changing nature of all things. Both anatta and anicca can be great sources of suffering, and not just because they entail the loss of all we hold dearly, but also because if we run away from this loss, we never get to experience the full beauty of caring deeply about something or someone to begin with. Like the poet in Freud's essay, if we cannot get comfortable (or at least comfortable enough) with the fact that everything changes, we risk going through life at an arm's length from its most poignant offerings. In trying to protect ourselves from the experience of change, we end up limiting the depth of our lives.

There is a saying in Buddhism that life is full of ten thousand joys and ten thousand sorrows. You don't get to experience the former without the latter.

In January 2015, the niche sport of rock climbing took over the world. Hundreds of reporters descended on Yosemite National Park. All the morning television shows were suddenly covering climbing, and so were the *New York Times* and *Wall Street Journal*. Two lanky and extremely fatigued men were approaching the summit of the Dawn Wall, a notoriously difficult route up a three-thousand-foot rock face reverentially known as El Capitan, or El Cap for short. Not only were the climbers ascending this challenging route, but they were doing it in free-form, a type of climbing that prohibits any artificial aids. Such a feat had never been accomplished, but not for lack of trying. Many of the best climbers in history had tried to free-climb the Dawn Wall and all had come up short. No one thought it was possible. The Dawn Wall was the last crown jewel of rock climbing. The big kahuna. Untouchable.

When you free-climb, all you've got at your disposal is a bag of chalk, some ropes to catch you if you fall, and your ten toes and ten fingers—or in Caldwell's case, nine. For the better part of three weeks, Caldwell led his partner, Kevin Jorgeson, up the rock. They'd eat. They'd climb. They'd sleep in a portaledge exposed to all the elements. And then they'd do it again. It was, by far, the most arduous climbing feat anyone could attempt in all of Yosemite, perhaps all of America or even the world. "This is, like, the hardest thing you could ever do on your fingers, climbing this route," Caldwell says in the documentary *The Dawn Wall*. "It's just grabbing razor blades." Yet Caldwell didn't seem to mind. He was climbing with a ferocious intensity, a hard-earned wisdom that emerged from years of life experience and struggle, and not just in climbing.

On Wednesday, January 14, at 3:25 p.m. Pacific Time, after nineteen days on the wall, the duo reached the summit, making history that reverberated around the world. The *New York Times* put it best, covering the story under the headline, "Pursuing the Impossible, and Coming Out on Top." As the sun set over the peak of El Cap, Caldwell was completely zoned in, immersed in the moment, seizing the stunning view and spectacular feeling, knowing that they, too, would soon pass. The highs, the lows, and everything in between all shared at least one thing in common: change.

In his book *Death*, the philosopher Todd May argues that if somehow you could become immortal, your life would not have as much meaning. It is a tough idea to fully grasp, one I struggle with myself. I get that if we lived forever, eventually we may get bored or it may start to feel like little is at stake. But Earth is a vast place, and who knows what the future of intergalactic travel holds. I think it would take quite a while, at least a few thousand years, before life would become a perpetual slog. So instead of focusing on immortality, let's imagine we could live for a few thousand years. That certainly sounds like a good deal. But even then, so long as we were still made of flesh and bone, we would be prone to tragedies, such as car accidents or contagions, that could end our lives. In order to ensure our full longevity, we would have to be increasingly cautious, perhaps so cautious that we wouldn't be doing much living at all.

To live is to lose. And it is the certainty of loss that makes life meaningful. What is change if not loss? The loss of youthful innocence. The loss of a finger. The loss of a job. The loss of a plan. The loss of a friend. The loss of a lover. The loss of how things were. The loss of how you thought things would be. When you first ponder change in this way, it may fill you with unease. I have been long reflecting on this topic while writing this book and it *still* deeply upsets me from time to time. This is especially true when realizing how fast my oldest child is growing up. Where did the time go? Can I please hit pause? I can't. This makes me sad. I feel tears well up behind my eyes.

But futile resistance or superficial delusion is no way to live. Yes, we will inevitably experience profound sadness at the reality of loss, and perhaps at times feel that our lives are terrifyingly small and insignificant when held against the infinite backdrop of the forever-changing universe. But we will also experience immense gratitude for all the wonder we find and for the wild and unlikely fact that we are here in the first place. Like an explorer on a path, the closer and more intimate we get with the changing landscape, the more beautiful, interesting, and deeply fulfilling the journey becomes, not in spite of the fact that we know it will change but, as Freud so eloquently pointed out, because of it.

The first core quality of rugged flexibility is opening to the flow of life. This doesn't mean that navigating change and impermanence will be easy. But we can learn to set appropriate expectations and develop the concrete skills to ready our minds and bodies for the challenge.

Open to the Flow of Life

- Embrace non-dual thinking: not this *or* that but this *and* that.
- Denying change may feel better in the short run, but it almost always feels worse in the long run, limiting the depth, texture, and potential for genuine excellence in one's life.
- Many of our problems, both individually and societally, result from resisting change.
- Only once we open to the flow of life and arrive at a genuine acceptance of change can things start to fall into place, empowering us to proceed pragmatically and productively on our respective paths.
- There are immense benefits to adopting a *being* over a *having* orientation; you become more rugged and flexible and less fragile to change. The things that you own no longer own you.
- If you catch yourself butting up against a wall, consider the inescapability trigger: What would it look like to fully accept your reality as it is? How might you work with it differently?
- Without change, our existence would become tedious and boring. If we are to live meaningful lives, change is simply part of the deal.

2

Expect It to Be Hard

In May 2021, after fifteen months of shutdown, quarantine, illness, and death, finally there was some light at the end of the tunnel. COVID cases in the United States had plummeted almost as quickly as they had once risen. The precipitous decline owed itself to a combination of vaccines, behavior change, herd immunity, and warm weather. In many parts of the country, your chances of getting into a car accident were higher than your chances of contracting the virus. Finally, after more than a year, people were going into the homes of family, neighbors, and friends without much, if any, worry. I distinctly remember the excited smile on my then three-year-old son's face when, for the first time in his living memory, one of his friends came over to play inside our home. "People can really come into our house, in real life!" he exclaimed, in contrast to his lived experience of social connection occurring exclusively outdoors or over FaceTime and Zoom.

My wife's anxiety about older and immunosuppressed family members decreased to baseline. I felt overjoyed for my biological brother and my best-friend brother, both of whom are big-city physicians and both of whom were utterly burned out from nonstop physical and emotional labor. My book *The Practice of Groundedness* was coming out that fall, and I began to look forward to the possibility of live events at my favorite bookshops. Like so many people, I was glad for the return of basic normalcy—we had waited long enough, or so we thought.

In early July cases began climbing again. The majority were identified as a newer, more contagious, and potentially more severe Delta variant. By the beginning of August 2021, infection rates were rising as fast as they had at any point during the pandemic. Whatever glimpse of normalcy had been bestowed upon us was shorn away as rapidly as it had come. The Delta variant was a gut punch. People were more devastated than ever. This was understandable; but it wasn't entirely logical.

Don't get me wrong, the rise of the Delta variant was terrible news. But on the whole, for most people things were still objectively better than they had been at the beginning of the pandemic. Vaccines, scientific miracles that reduced hospitalization and death rates on the order of ten to twenty times, were widely available. New therapeutics began hitting the market. Public health knowledge about contagion pathways and subsequent mitigation strategies had increased substantially. Nonetheless, the Delta variant of COVID was met by omnipresent despair. Everyone was expecting one thing—the gradual unwinding of the pandemic—but what we got was something else entirely.

How Expectations Alter Reality

Imagine it is nearing the end of a long day's manual labor during which you ate nothing. You're starving. You'd eat anything, but if you could pick, you'd have your favorite food, spaghetti Alfredo. Around five in the afternoon, someone interrupts your work and explains that, come six thirty, a large plate of the dish will be waiting for you, prepared by no less than a Michelin-starred chef. Saliva fills your mouth. Perhaps you even pucker a bit. If your hunger level had been at a nine, now it is a ten. Focusing on your work becomes hard, but you do your best to return to it and finish for the day. Six thirty arrives. You are feeling famished as you're escorted to a room with a dining table. Out of the back comes an irritating person from your neighborhood, Billy. He is the kind of man who grumbles for no apparent reason as he passes you on the sidewalk, his grumbling dog always in tow. Billy is holding a large bowl of unsalted and slightly stale pretzels. Billy informs you that the spaghetti you were promised was a cruel joke he played on you with his cousin, the person who told you about it earlier. He proceeds to leave the room, mumbling, *"Bon appétit"* on the way out. How do you think you'd feel? When posed with this question, most people respond that they'd be infuriated, though their situation is markedly better than it was before. After all, starving with food beats starving without food, even if what's on offer are some slightly stale pretzels.

Swaths of psychological research show that our happiness in any given moment is a function of our reality minus our expectations. When reality matches or exceeds expectations, we feel good. When reality falls short of expectations, we feel bad. Countries that consistently rank as the happiest are not necessarily better than their neighbors. But the citizens in these countries tend to have lower expectations. In a landmark study, epidemiologists from the University of Southern Denmark set out to explore why their citizens regularly score higher than any other Western country on measures of happiness and life satisfaction. Their findings, which were published in the *British Medical Journal,* focused on the importance of expectations.

"If expectations are unrealistically high they could be the basis of disappointment and low life satisfaction," write the authors. "While the Danes are very satisfied, their expectations are rather low." This is in stark contrast to so many other places in the Western world, where, starting at a young age, people are raised to believe that a hedonic version of happiness is the ultimate goal and they should expect it always.

A key characteristic that separates allostasis, the new and more accurate model of change, from homeostasis, the old model, is that allostasis has an anticipatory component. Whereas homeostasis is agnostic to expectations, allostasis states that if you expect something to happen you will suffer less distress during the ensuing period of disorder. For example, homeostasis says that it doesn't matter if you get shot in the leg on the battlefield or shot in the leg at the grocery store, your response will be the same—you got shot in the leg. Allostasis more accurately recognizes that the responses will be different. The person who gets shot in the leg on the battlefield will experience less psychological and even physiological distress, a phenomenon that can be observed all the way down to the hormones circulating in one's blood. Unlike the shopper at a grocery store, in the soldier's mind getting shot in the leg was a possibility, if not an expectation.

It follows that a crucial part of rugged flexibility is setting appropriate expectations. We'll continue this chapter by further exploring why expectations are so important, delving into fascinating and cutting-edge neuroscience. Then we'll discuss three powerful, evidence-based, and concrete methods to set appropriate expectations in a way that protects us from blind optimism and toxic positivity on the one hand, and doom and despair on the other. We'll also explore how pain and suffering are two similar yet different phenomena, and examine the pliable relationship between them.

The Brain Is a Prediction Machine (The Neuroscience of Expectations)

The reason that expectations affect us so dramatically traces itself to the neural circuits that connect your prefrontal cortex (the thinking part of your brain that controls voluntary action) with your more ancient brain stem (the feeling part of your brain that controls involuntary action). Up until a few decades ago, the predominant view in neuroscience held that consciousness was mainly your brain experiencing the world as it is. More recent research—spearheaded by the neuroscientists Andy Clark at the University of Edinburgh, in Scotland; Jakob Hohwy at Monash University in Melbourne, Australia; and Mark Solms at the University of Cape Town, in South Africa—shows that your brain functions more like a prediction machine. Your prefrontal cortex is constantly generating predictions for what might unfold. These predictions are sent to the brain stem, which then prepares your mind-body system for whatever is anticipated. The brain adopts this forward-looking stance for good reason: it is far more efficient than approaching every moment with no notion or bias for what might happen next. Imagine you are at an airport, walking through the jet bridge about to board a plane. Without its predictive function, the brain would have to be equally prepared for you to walk off a cliff, into a pool, or into traffic. This would be horribly inefficient and consume all of your neurological energy. In our evolutionary past, it would have amounted to a huge survival disadvantage. Today, you'd simply never get anything done.

"The prediction component brings out the idea that neural subsystems operate not just on the basis of actual signals from communicating subsystems, but on their dynamic predictions of such signals in hierarchically organized networks. Such cascades of multiple predictions introduce the need for regulatory processes, both local and supervening, which handle error signals, assign signal weights, as well as influencing gain-modulation in other parts of the system, in pursuit of stable and energy-efficient processing," write a team of neuroscientists from the University of Gothenburg, in Sweden, in the journal *Frontiers in Human Neuroscience.* In layperson's terms, the brain begins with an expected scenario that is continually adjusted to match reality; the closer the match, the better we feel and the less energy we burn.

In a famous experiment led by Nobel Prize–winning psychologist Daniel Kahneman, researchers instructed participants to submerge their hands in painfully cold water for sixty seconds, and then again for sixty seconds plus an additional thirty seconds, during which time the water was warmed from fourteen degrees Celsius to fifteen degrees Celsius. When asked which trial they would like to repeat, the vast majority of participants said the second one, even though the total discomfort was greater: sixty seconds of super cold water plus thirty seconds of quite cold water is worse than just sixty seconds of super cold water. But it was the fact that the conditions gradually got better toward the end of the second condition that led participants to favor it. Kahneman and colleagues replicated this finding in diverse settings. For example, most people rate more positively an experience in which they wait in a slow line for forty-five minutes and then the line speeds up in the last ten minutes versus just waiting in a slow line for forty-five minutes, even though the total wait time is longer in the first condition.

When these studies were first published in the mid-1990s, the main implication was that people place inordinate value on the end of an experience. Given what we now know about consciousness and the brain's anticipatory function, I suspect the mechanism behind this is that during any given experience, we develop an expectation of what will happen next. At the end of an experience, that expectation is either met (the cold water stays the same; so does the pace of the line), not met (the cold water gets even colder; the line slows down), or exceeded (the cold water warms up; the line moves faster). We subjectively prefer when our expectations are exceeded, even if that means more total units of objective distress. It is a theme that will come up repeatedly in this chapter: *consciousness is not solely our experience of reality; it is our experience of reality filtered and modulated by our expectations for it.*

More recent studies demonstrate that expectations don't just influence our perception of current experiences and remembrance of past ones; they also affect how we approach the future in a multitude of ways. On its face this sounds obvious, but the implications run deep. For instance, when fatigued and crashing athletes are told they are near the finish line of a race, they start to feel better and mysteriously find a second wind—presumably because their brains, anticipating the finish line around the corner, stop conserving energy, allowing them to empty their tanks. Other experiments show that when exhausted athletes rinse their mouths with a sports drink, they immediately feel better and generate more power. There's only one catch: after rinsing their mouths, they spit out the sports drink. The calories and nutrients never make it down the athletes' throats, let alone into their stomachs. Presumably, the brain gets a taste of the sports drink, predicts it will soon be digested, and subsequently loosens its grips on the body's ability to push itself. However, if over time the athletes repeatedly spit out the sports drink, it would cease to have a positive effect. The brain would stop associating sports drink in the mouth with oncoming calories. The same is true for the athletes who were told the finish line is nearby when it was not. Once their brains figured out what was going on, the participants would be unable to speed up. You can only fool the brain for so long.

That the brain is a prediction machine is, on balance, hugely advantageous. However, in cases where expectations don't match reality, we get thrown for a loop. The bigger the mismatch, especially if expectations are rosier than reality, the worse the suffering. This is true not only psychologically but physiologically too. Remember that a bad prediction requires more energy to get back in sync with reality. The more out of sync our expectations are, the more problematic, since our brains and bodies are built to conserve energy. What you and I experience as consciousness is, in many ways, the ever-present cascade of thoughts and feelings our brains generate to let us know whether or not our predictions are on the right track. If our predictions are accurate, we feel good and have generally calm and happy thoughts. If our predictions are overly optimistic, however, we feel bad and our thinking turns negative.

Herein lies the fascinating connection between our basic psychology, or what many would call our "minds," and our basic biology, or what many would call our "brains." The psychological equation that says "happiness equals reality minus expectations" essentially represents the accuracy of our biological—that is, our brains'—predictions. When I first realized this I was mind-blown, and for good reason. I had no idea that the unification of psychology and biology was where my research would lead. It was an insight I could have never predicted, hence the associated strong feelings.

Rewind to the summer of 2021 and the emergence of the COVID Delta variant. Even if most people were empirically better off than at the start of the pandemic, in light of what we just covered, of course everyone felt devastated. It was as if someone had told us we were at mile twenty-four of a marathon, and then right as we were emptying out all of our reserves and hitting our finishing stride, they dropped us back at mile eleven. In the fall of that same year I wrote a short piece making this exact analogy. An astute reader responded: "This is spot-on, with one glaring exception: COVID is not like a marathon; it's more like an ultramarathon trail race in which we don't know where the finish line is."

His thoughtful comment goes beyond COVID. Remember from the introduction of this book that the average adult experiences thirty-six major life disruptions. In many ways, the entirety of our existence is like an ultramarathon in which we know neither where the finish line is, nor what obstacles will pop up along the way. The glaring question, then, is how on earth are we supposed to run the race?

Tragic Optimism

Serge Hollerbach was born in Leningrad, Russia, in 1923. At seventeen, he enrolled in a high school for creatives. Six months after he enrolled, in June 1941, German forces invaded. They forced him and many other Russians to work as laborers in Nazi factories. He survived the internment, and once the war ended, he enrolled in the Munich Academy of Fine Arts. World War II shook Hollerbach to his core, eviscerating any naive innocence he might have once had. Yet he remained optimistic about life, in large part due to his love of craft, which represented reality but also gave him an opportunity to transcend it. He found joy and meaning in creative expression.

At the Munich Academy of Fine Arts, Hollerbach was schooled in expressionism, a visual style which presents the world from a subjective viewpoint, attempting to reproduce how one experiences reality rather than reality itself. Famous examples of expressionism include Vincent van Gogh's *Starry Night* and Edvard Munch's *The Scream*. In his later years, Hollerbach would lean heavily upon his expressionistic education, but for reasons no one could have foreseen.

In 1949, Hollerbach immigrated to the United States and settled in New York City. There, he hit his stride as an artist, producing a number of casein and watercolor paintings that were featured in several museum collections, including the Yale University Art Gallery, the Butler Institute of American Art, and the Georgia Museum of Art. Hollerbach became known for a unique approach to painting that combined expressionist techniques with bold and realist sensibilities to convey the essence of the human experience. His work received many accolades, including the American Watercolor Society's Gold Medal in 1983 and Silver Medals in 1989 and 1990, the Audubon Artists Silver Medal in 1983, the Grumbacher Gold Medallion in 1988, the Allied Artists of America Gold Medal in 1985 and 1987, and first prize in the Rocky Mountain National Watermedia exhibition in 1986 and 1987. In addition to his exhibitions and commercial work, he also taught at the National Academy of Design.

In 1994, at age seventy-one, Hollerbach's vision began deteriorating. Shortly thereafter he was diagnosed with macular degeneration, a disease that typically affects elderly adults, destroying their central vision, degrading their peripheral vision, and leaving most legally blind. Hollerbach's decline was rapid. Although eventually surgery stabilized his vision, the disease had taken a severe toll, leaving Hollerbach with no central vision and unable to make out any details—all he could see were big and general shapes. "Well, I see your face, but I have to come very close to you. But I couldn't do a portrait, I couldn't sketch you. I see everything in a blur. It's out of focus," he recalled. Hollerbach, the masterful visual artist who made his name perceiving fine details no one else could, was pronounced legally blind.

He was incredibly frustrated to have lost his ability to perceive intricacies, especially given how much he enjoyed painting human subjects. But he also realized there was nothing he could do to regain his vision. Any resistance would be futile. Instead of despairing, he decided to emphasize the expressionist quality of his work and let the realism go. He swiftly moved away from attempting to portray what he saw in front of him and toward painting what he saw within him, what he called relying on his "inner eye." This, he said, allowed him to convey "what is most important in life."

Rather than give up on painting when his vision abruptly declined, Hollerbach adapted his relationship to the craft, eventually landing on something that felt perhaps even more authentic than his former realistic approach. Hollerbach didn't view the loss of his vision as positive. But he didn't see it as unambiguously negative, either. "It's a very sad thing, but not a major catastrophe," he explained. "In a way, my vision impairment gave me new direction. I wouldn't say it's a blessing. But it gave me a new venue. I think that my visual impairment led me back to what I could have been."

The term "tragic optimism" was coined by another World War II survivor, Viktor Frankl, a Jewish psychologist from Vienna who endured the Nazi death camps. Frankl is well-known for his book *Man's Search for Meaning*, first published in 1946. *Man's Search for Meaning* is part Holocaust memoir and part psychology text. The second half of the book builds the foundation for what became existential psychotherapy, Frankl's system for finding fulfillment and meaning, even in the direst of circumstances. It has been translated into more than fifty languages, sold over sixteen million copies, and is considered required reading for any student of human nature.

What most people don't know is that in the mid-1980s Frankl wrote a postscript to the book, a short essay he titled "The Case for Tragic Optimism." In it, Frankl observes that life involves three inevitable varieties of tragedy: the first is pain and suffering, because we are made of flesh and bone; the second is guilt, because we have some freedom to make choices and thus we feel responsible when things don't work out as we had hoped; and the third is our ability to look ahead, because we must face the fact that everything we cherish, including our own lives, will eventually change or end. Even though we live with these three inevitable varieties of suffering, Western society places immense pressure on everyone to be relentlessly happy. At best this is misguided; at worst it is dangerous. As you've read, too-rosy expectations are a common cause of disappointment and distress. Meanwhile, judging yourself for feeling down, or internalizing the idea that there is something wrong with you when you are sad (or that sadness is a weakness), only makes whatever you are going through harder.

In my own experience, the worst way to be happy is by trying to be happy all the time, or worse yet, assuming (and expecting) that you ought to be. I suspect many people don't realize the heavy emotional burden they carry by absorbing an ethos that says, implicitly and sometimes even explicitly, that you should be positive and upbeat always—even though sadness, boredom, and apathy are inevitable parts of the human experience. I also suspect that much of the judgment we levy upon ourselves, and the impatience and lashing out we do at others, comes from carrying the weight of this impossible standard. In a study including over seventy-thousand individuals from all over the world that was published in 2022 in the *Journal of Personality and Social Psychology*, researchers found that people's experience of happiness and fulfillment are correlated to the accuracy of their expectations. Instead of placing unabashed happiness on a pedestal and making it our primary goal, perhaps we ought to embrace tragic optimism as a better alternative.

First, a definition: tragic optimism is the ability to maintain hope and find meaning in life despite its inescapable pain, loss, and suffering. It is about acknowledging, accepting, and expecting that life will contain hardship, that sometimes impermanence hurts, and then trudging forward with a positive attitude nonetheless. With tragic optimism, if a situation doesn't unfold as badly as you thought, you'll be pleasantly surprised. If a situation does unfold as badly as you thought, you'll be prepared and levelheaded. Research shows that individuals who face life with a mindset of tragic optimism—in particular, those who expect a fair share of change and hardship—have advantageous physical and psychological responses to stress. They feel less pain, gain more fortitude, and are more likely to successfully move forward following disruption. Just think about how many times a toddler falls while learning to walk or run. They may get bumps and bruises, but they certainly don't feel as much pain or get as discouraged as would an adult. At that stage of development, toddlers don't expect anything less than a massive struggle, and thus they are ready to confront it.

It is worth being explicit that tragic optimism is not about actively seeking out suffering. I feel strongly that if we can avoid suffering we should. Rather, tragic optimism is about realizing the inevitability of suffering, that life gives us plenty of practice on its own, and also that we generally have at least some say in how we face it. "Is this to say that suffering is indispensable to the discovery of meaning? In no way. I only insist that meaning is available in spite of—nay, even through—suffering, provided that the suffering is unavoidable," writes Frankl. "If it is avoidable, the meaningful thing to do is to remove its cause, for unnecessary suffering is masochistic rather than heroic. If, on the other hand, one cannot change a situation that causes his suffering, he can still choose his attitude." This, of course, is exactly what Frankl did as a survivor of the Holocaust, and what Hollerbach did in response to his blindness.

A Scientific Theory for Why Tragic Optimism Works

Frankl's work on tragic optimism preceded the latest neuroscience on the predictive brain. But knowing what we know now, I want to put forth an argument for why tragic optimism is such an effective outlook. If you expect and predict life to be hard, then you won't be surprised when it is—which in and of itself makes life easier, and also improves your chances of finding equanimity and meaning amidst change and struggle. Another powerful example of non-dual thinking, tragic optimism teaches us that life can be sad *and* meaningful, that we can experience pain *and* joy, that change can bring about anguish *and* hope, and that impermanence represents both endings *and* beginnings. If nothing else, tragic optimism is a more accurate way to conceptualize a messy world, one that is full of complexity and contradictions. And, as we've learned, the brain favors accurate conceptions and expectations.

Also inherent to tragic optimism is an acceptance of whatever emotions you feel in response to change and disorder. For example, following the terrorist attacks on September 11, many people understandably reported increased feelings of fear, anxiety, depression, dread, and despair. But these emotions were more debilitating and persistent for some than for others. A team of researchers from the University of North Carolina at Chapel Hill and the University of Michigan in Ann Arbor set out to learn why. They found that the more resilient people fully acknowledged and felt the horror of what happened. They experienced the same levels of sadness, stress, and grief as the less resilient people, but they were able to hold room for emotions like love and gratitude too.

This study is one of many showing that tragic optimism is a useful quality not because it numbs pain or turns you into a Pollyanna, but because it widens your inner aperture, creating space for you to hold an expansive range of feelings—which is an "accurate expectation" for what it means to be human. Tragic optimism says that you can still enjoy a stroll through the woods on the same day something terrible happens in the world. It also says that you can be sad and down even though there may be a lot that is good in your life. On many days, you'll feel all of these emotions—not because something is wrong with you but precisely the opposite: because all of these emotions are part of even the most average human existence. What shifts with tragic optimism is that all the usual repression, delusion, self-judgment, rumination, and despair get left behind. This outlook opens up the space for you to commit to wise hope and confront your circumstances with wise action, concepts to which we'll turn next.

Wise Hope and Wise Action

At the heart of Buddhist psychology lie a few ancient Pali and Sanskrit texts. In them, the word *dukkha* surfaces repeatedly.

The first noble truth of Buddhism, the premise that underlies the entire philosophy, is *dukkha-satya,* or "the truth of dukkha." Today, the word *dukkha* is commonly translated into "suffering." But this translation is not exactly accurate. *Du* is the prefix for "difficult" or "hard," and *kha* has many meanings, including "to face." Put them together and what you get is that *dukkha* actually means "hard to face." Unlike what so many people think, the first noble truth of Buddhism does not teach that life is suffering; rather, it teaches that life is full of things that are hard to face. Perhaps suffering is the most common byproduct of dukkha, but it is not the thing itself.

That life is full of things that are difficult to face was true in the Buddha's era 2,500 years ago and remains true today. Examples include personal injury or illness, climate change, threats to democracy, a global pandemic, and age-related decline, to name just a few. In the face of all this dukkha, two attitudes tend to prevail. Some people choose to bury their heads in the sand, delude themselves, or express a toxic positivity. Others choose to be excessively pessimistic or despairing. Both of these attitudes are easy to adopt because they absolve you of doing anything. The former denies that anything is wrong; and if nothing is wrong, there is nothing to worry about, nothing to change. The latter takes such a grim stance that it basically says any action would be pointless, so why bother—it is a fast track to helplessness and nihilism. Neither of these attitudes are particularly helpful. But somewhere in between exists a third way, an approach that is a natural extension of tragic optimism: committing to wise hope and wise action.

Wise hope and wise action ask that you accept and see a situation clearly for what it is, and then, with the hopeful attitude necessary, say, *Well, this is what is happening now, so I will focus on what I can control, try not to obsess over what I can't, and do the best I can. I've faced other challenges and other seasons of doubt and despair, and I've come out the other side.*

Wise hope and wise action are not just pathways to productively engage in and influence change and disorder. They also support mental and physical health. Hopelessness and helplessness are associated with clinical depression and physical decline. Meanwhile, toxic positivity is associated with increasing levels of the stress hormone cortisol, which leads to high blood pressure, headaches, insomnia, obesity, and many other modern maladies (because delusion requires a lot of work). If, however, we can respond skillfully to changes that are hard to face, if we can respond with wise hope and wise action, then we diminish our maladaptive reactions and become more resilient as a result.

In 1985, a young man named Bryan Stevenson graduated from Harvard with both a master's in public policy and a juris doctorate from the school of law. His attraction to the law had always been rooted in the protection of civil rights and the provision of equitable justice. He became a staff attorney at the Southern Center for Human Rights in Atlanta, Georgia, an organization that represents capital defendants and death row prisoners. In 1989, after a few years of working with convicts in the Deep South, Stevenson founded the Equal Justice Initiative (EJI), a human rights organization, in Montgomery, Alabama, dedicated to ending mass incarceration and protecting the rights of the most vulnerable people in America, such as those on death row. Wherever Stevenson looked, he saw unfairness, brutality, and suffering. Trying to provide a counterforce became his life's work.

Over the past three decades, Stevenson and the EJI have won major legal challenges eliminating excessive and unfair sentences, exonerating innocent death row prisoners, confronting the abuse of the incarcerated and mentally ill, and aiding children being prosecuted as adults. Stevenson has argued and won multiple cases at the United States Supreme Court, including a landmark 2012 ruling that banned mandatory life imprisonment without parole for children seventeen or younger at the time of their conviction. He has won reversals, relief, or release for more than 135 wrongly condemned prisoners on death row, and he has helped hundreds more. For his work representing heavily disadvantaged and disenfranchised populations, Stevenson has won the MacArthur Foundation's genius prize and countless other awards and accolades. His 2014 memoir, *Just Mercy,* became an immediate bestseller and was turned into a movie by the same name, in which he was portrayed by the actor Michael B. Jordan. In addition to his client work and advocacy, Stevenson is also a professor at the New York University School of Law, and he's spearheaded the creation of two cultural heritage sites in Montgomery, Alabama, both of which are dedicated to the relationship between slavery, segregation, and mass incarceration.

Stevenson has had as productive a career as anyone. His accomplishments would be extraordinary in any field, but they are especially remarkable given his line of work and the daunting challenges he faces. Representing capital defendants places you at the center of unimaginable pain and suffering. There are the falsely accused and the rightfully accused; the families of the defendants and the families of the victims; and the countless examples of racism and other forms of discrimination in the system. As such, for the protection of both their physical and emotional health, many public defenders keep a safe distance from their clients. But Stevenson operates differently. He gets close.

"I believe that we are all more than the worst thing we have ever done. I don't think that if someone tells a lie they are just a liar. I don't think that if someone even kills somebody, they are just a killer. And I think that justice requires us to see the other things that you are. And if the advocate can't get close enough to see what those things are, you're not going to do a very good job," says Stevenson, who is known for going into jails and prisons and spending hours upon hours sitting with defendants, hearing their stories, affirming their humanity, and offering them dignity.

Stevenson is under no delusion about the stark problems in our modern justice system. He is also under no delusion that some of the people he represents are guilty of heinous crimes, the majority of which arise out of heinous conditions. His proximity to dukkha upends any chance of naivete or willful blindless. Yet he doesn't fall into nihilism or chronic despair. Epitomizing wise hope and wise action, Stevenson says that "ultimately, we are talking about a need to be more hopeful, more committed, and more dedicated to the basic challenges of living in a complex world." Innovation, creativity, and development come not from the ideas in our mind alone, he says, "they come from the ideas in our minds that are also fueled by some conviction in our hearts. And it's that mind-heart connection that I believe tells us to be attentive to not just all the bright and dazzling things, but also the dark and difficult things . . . There are definitely hard days, difficult days, painful days, but I'm really grateful I've seen justice rise up, I've seen truth prevail, and that's an amazing thing."

Stevenson's story is a profound example of wise hope and wise action, which is precisely why I've chosen to include it. If Stevenson can demonstrate wise hope and wise action under the conditions in which he operates, then so can we in our own lives. It is also worth pointing out how sad it is that the current justice system requires Stevenson's heroic work to begin with. But again, this is precisely why his story is included. Holding on to hope can be hard work, especially in the circumstances we need it most. Many things in our world are dysfunctional, there is seemingly endless dukkha. Facing it is challenging, but the alternative—doing nothing, be it because of willful blindness or despair—is undoubtedly worse. If we are to have any chance at improving a broken world, we must not become broken people.

Does this mean we all need to do the work that Stevenson does and at the level he does it? No. But it ought to inspire us to confront the difficulties in our own lives with wise hope and wise action, the former opening up the opportunity for the latter. When we expect life to be hard but also keep our minds and hearts receptive to joy and possibility, when we embrace tragic optimism and follow it with wise hope and wise action, we sturdy ourselves to walk our paths wherever they may take us, even if that means into the dungeons of death row.

Hope, writes the moral philosopher Kieran Setiya, "keeps the flicker of potential agency alive." Action is essentially impossible without hope; for there would be no reason to do anything absent at least some belief that it might yield fruitful outcomes. What makes wise hope and wise action so hard, then, is that they make us vulnerable to further loss and pain if things don't go our way. They force us to put our skin in the game. But then again, isn't that the whole point of being alive?

Suffering Equals Pain Times Resistance

Imagine that you have a sore lower back. On a scale of one to ten, you rate your pain as a six. Now imagine that you become very frustrated by the pain. You are upset that it is ruining your day and, worse yet, worried that you will not be able to hike with your friends on the coming weekend, as you had planned. It doesn't help that the ibuprofen and Tylenol you just took are not having any effect. Your worry quickly spirals into catastrophizing and you fear that the pain will never go away. You begin to believe that you may feel like this forever. In addition to the six units of pain you are experiencing, you've now appended seven units of resistance. Only resistance is not additive to pain; it is generally a multiplier. In other words, suffering is not the same thing as pain: suffering equals pain times resistance. In this example, you've got forty-two units of suffering—that is, six units of pain multiplied by seven units of resistance. The more you resist your pain, the exponentially worse your suffering becomes. Fortunately, the same math works in the opposite direction. Continuing with the example above, if you could lower your resistance to three units, your total suffering would drop to eighteen—that is, six units of pain multiplied by three units of resistance. Although this equation may not be mathematically perfect, research shows that it is conceptually accurate.

Consider the Mayo Clinic's world-class Pain Rehabilitation Center. People from all over the world travel to Rochester, Minnesota, to enroll, often as a last-ditch attempt to eliminate their suffering after everything else has failed. Patients arrive with a wide variety of difficulties including chronic back pain, fibromyalgia, headaches, neuropathy, chronic fatigue syndrome, and all manner of digestive disorders. The program relies on a multifaceted approach including physical therapy, cognitive therapy, behavioral therapy, biofeedback, and education. The ultimate goal of the program is not to eliminate patients' pain, but rather to eliminate the overwhelming desire of patients to eliminate their pain. This starts with helping patients taper off opioids and other medications, and ends with them learning to update their expectations about pain and to accept that some amount is okay, thus lessening their resistance. A core aim of the program is to de-catastrophize discomfort and gradually increase the number of activities in which patients are able to participate.

Cathy Jasper knows this firsthand. In her early sixties, she began experiencing strange symptoms that grew quickly in number and intensity. They included forgetfulness, weakness in the left side of her body, extreme back pain, and allodynia, or pain caused by activities that generally do not cause discomfort. The allodynia got so bad that it prevented Jasper from hugging her husband, putting her elbow on a table, and after a while, even eating. Her symptoms would last for about a month, go away for several months, and then return, seemingly at random. During a particularly rough patch, she also started having seizures, became debilitated, and lost twenty-eight pounds.

Jasper searched everywhere for an explanation of what was happening to her. She sought care from several doctors, underwent an evaluation at an epilepsy center, and started taking alternative medicines, supplements, and CBD oil in an effort to ease her discomfort. After two years of suffering in this way, she was diagnosed with central sensitization syndrome, a condition in which the central nervous system amplifies signals that are sent to the brain's sensory and motor cortices, triggering a wide range of disconcerting symptoms. After she received the diagnosis, she enrolled at the Mayo Clinic's Pain Rehabilitation Center.

A particularly effective part of the treatment for Jasper—and for so many others in the program—is called *graded exercise exposure*. In it, patients who are convinced that their pain prohibits them from participating in certain activities are gradually exposed to those activities, all under the supervision of therapists or physicians. In most cases, they realize that if they can work through the initial bolus of pain, they start to settle in and feel okay, if not better. Over time, they are exposed to greater and greater challenges. "Pain is a warning signal from your brain that you are afraid of injuring yourself," says David Brown, a physical therapist at the program. "But sometimes the brain erroneously sends these pain signals. This graded exercise approach retrains your brain to understand that it's safe to move," he says.

Brown explains that in Jasper's case, "She had a lot of verbal and facial expressions of pain. She not only avoided daily activities because of pain, she also had withdrawn from social activities due to the spells she was having." A huge part of Brown's job is to help patients like Jasper identify their habitual pain behaviors and patterns of resistance and then come up with strategies to stop them. The goal is not to eliminate pain and other symptoms but rather to help patients face their pain and discomfort more skillfully so they can live fulfilling lives, and in doing so, minimize their overall suffering.

By the end of her treatment, Jasper went from having ten spells of symptoms per day to none. She could walk more than one third of a mile in six minutes, about a 20 percent increase from when she started the program. And the effects have been lasting. "Nearly two years later, I am continuing to do cardio and physical therapy three to five days a week. The physical therapy keeps me balanced," says Jasper. "I'm able to participate in and keep up with group conversations. I'm sleeping eight hours a night," she says. "I can take care of my sixteen-month-old grandson and be a caregiver for a relative with Alzheimer's who lives with us. My husband travels internationally for work, and I can go with him." Though Jasper still experiences pain and other symptoms, her suffering is far less than it was, namely because she has shed so much of her resistance. This is not to say that pain isn't real and cannot be debilitating. It is simply to say that many people find benefit in learning to lessen their resistance, hard as it may be.

The Mayo Clinic's Pain Rehabilitation Center is startlingly effective for two reasons: First, it shifts patients' expectations around pain from something to be avoided and cured to something to manage. Second, it teaches patients to lessen their resistance. It is an approach that exemplifies the two important equations in this chapter: happiness equals reality minus expectations, and suffering equals pain times resistance. If you can align your expectations with reality and minimize your resistance to pain and discomfort—or more broadly, to everything that is hard to face, to the truth of dukkha—you set yourself up for the best experience and outcome, regardless of what you are facing. Remember that the brain is a prediction machine. You may want life to go a certain way, plan for it, and do everything you can to manifest those plans, but at some point things inevitably go awry. The more you pull back and withdraw, which is its own form of resistance, the worse off you'll be. The crucial work is updating your expectations and confronting reality, even if doing so feels difficult and uncomfortable at first.

Whether it's a condition like central sensitization syndrome, the emergence of a new virus variant, or a much lesser change, the more quickly you are able to let go of futile resistance and skillfully face what is happening, the better you feel and the more you find yourself able to do. Tragic optimism helps you set appropriate expectations. Wise hope and wise action allow you to move forward with grace and grit.

Our paths will serve up all manner of difficulties. That's just how it goes. All we can do is call a spade a spade—even, and perhaps especially, if we thought we'd receive a diamond—and then work with the hand we've been dealt.

A Rugged and Flexible Mindset

The two core components of a rugged and flexible mindset work together. First, we've got to drop the weight of denial and resistance and instead *open to the flow of life,* accepting that the only constant is change and seeing it clearly for what it is. Second, we've got to *expect it to be hard,* which, paradoxically, makes everything easier. As you've read, these mindset shifts are powerful because our experience of impermanence, and thus our ability to work with it, is contingent upon how we view it. The goal of adopting a rugged and flexible mindset is to enhance that view—or in neuroscience speak, to enhance our predictions—by making it more nuanced, complex, and accurate. When we shed our preconceived biases and delusions, we feel and do better. A rugged and flexible mindset serves as the foundation upon which we can build a new, advantageous, and freer relationship with change and disorder, one that skillfully navigates the inevitable obstacles and undulations on our paths, and even grows from them.

Many of the people whose stories you read in the first two chapters—whether it's Tommy Caldwell, my client Christine, Serge Hollerbach, Cathy Jasper, or Bryan Stevenson—underwent not only external change but internal change, too. On the one hand, these people are the same as they've always been. On the other, they evolved dramatically throughout their lives and as a result of their experiences. This is true not just for them, but for all of us. As we walk our respective paths and navigate ongoing cycles of order, disorder, and reorder we'll let go of certain qualities, characteristics, and attitudes that we've been carrying, and we'll pick up new ones to bring along the way.

What, then, does it mean to have a strong and enduring identity when everything, including ourselves, is always changing? How do we create a sense of self that is both rugged and flexible, that can withstand and grow from change? These are the topics to which we'll turn in the next part of this book.

Expect It to Be Hard

- A key characteristic that separates allostasis, the new and more accurate model of change, from homeostasis, the old model, is that allostasis has an anticipatory component: whereas homeostasis is agnostic to expectations, allostasis states that expectations shape our experience.
- Happiness at any given moment is a function of your reality minus your expectations.
- Our culture pushes us to wear rose-tinted glasses and "think positive," but we have a better chance at feeling and doing good if we set realistic expectations—including that things change all the time, sometimes for better and sometimes for worse.
- Our brains are constantly trying to predict what will happen next and then align those predictions with reality—when our predictions are off, we benefit from updating them as swiftly as we can.
- There are numerous advantages to cultivating an outlook of tragic optimism, realizing that life contains inevitable pain and suffering yet moving forward with grace and grit nonetheless.
- When confronted with significant challenges, instead of being a Pollyanna or wallowing in despair and nihilism, both of which are maladaptive, do what you can to commit to wise hope and wise action. *Like it or not, this is what is happening right now; I am going to focus on what I can control, do the best I can, and come out the other side.*
- Suffering equals pain times resistance; the more you can shed your resistance, the exponentially better you'll feel and do.

Part 2

RUGGED AND FLEXIBLE IDENTITY

3

Cultivate a Fluid Sense of Self

I distinctly remember waking up on a chilly February morning in 2022, my phone full of messages: *Have you seen the Nils van der Poel document?*

Shortly after winning two gold medals and posting a world record at that year's winter Olympic Games in Beijing, China, the twenty-five-year-old Swedish speed skater Nils van der Poel did something nobody was expecting. He published a sixty-two-page PDF entitled "How to Skate a 10K . . . and Also Half a 10K," referring to the two events he won at the Olympics. Rarely, if ever, do world-class athletes share their training programs, which amount to proprietary, top-secret formulas. For that reason alone, van der Poel's decision to make his training public was interesting. But I was still confused as to why so many people thought of me when they read the document. I am not a speed skater, and I don't follow the sport closely. My curiosity was piqued. I went over to my computer and clicked on a link to download the PDF. Within a few minutes, I knew I was in for a treat.

The first page of this so-called "training" document was

entirely blank but for a single quote from the psychologist Carl Jung: "It seems that all true things must change and only that which changes remains true." In February 2022, I had not yet shared the idea for the book you are currently reading with anyone outside of my wife, my literary agent, and my editor. So you can imagine the chills that shot down my spine when I read that. Yes, the PDF contained all sorts of specific workouts and exercise protocols. But it also contained an exposition on the pursuit, meaning, and value of excellence—trials, tribulations, and disruptions included. And that, dear readers, is the content I am here for.

It is widely known that endurance athletes make some of the best philosophers. Running, skating, cycling, and swimming are all solitary endeavors. Anyone who takes these pursuits seriously ends up spending a whole lot of time in their own head. Van der Poel, who trained for upward of seven hours per day during his preparation for the Olympics, was no exception. In those seven hours, he reflected extensively on his identity and self-worth.

Whereas many Olympians define themselves by their sport, designing every hour of their lives around it, van der Poel did not. In the buildup to the 2022 Olympic Games, instead of using his rest and recovery days to lie on the couch drinking protein shakes, receiving copious amounts of massage work, and sleeping—which is what just about every other world-class athlete does—he went out with friends. "My rest days were usually during weekends," he writes. "In that way I could spend the weekends doing fun stuff with my friends. Usually I did not train at all during rest days. I rested both my mind and my body. However, if my friends wanted to go alpine skiing or go for a hike, I would join them. But I didn't perform any [specific] recovery. I tried to live a normal life . . . I drank beers like any other twenty-five-year-old." For an athlete of van der Poel's caliber to sacrifice two days a week to normalcy in a buildup to the Olympic games is unprecedented.

Van der Poel wasn't always like this. In his youth he identified fully with speed skating and its culture, becoming dependent upon his success in the oval. "As a teenager the sport meant everything to me, which I do not believe is a good thing," he explains. When training and competition were going well, he was elated. But something as trivial as a poor workout could send him into a steep and downward spiral. After a few years of riding this emotional roller coaster, one that is common to highly driven individuals in pretty much every field, van der Poel decided it was an unsustainable way to train, let alone live. He could not be just a speed skater. The sport might make up a significant part of his identity, but it could not comprise its entirety.

And so, in his early twenties, van der Poel began focusing on building a life outside of sport. He went out for pizza and beer with friends who had nothing to do with speed skating, and he read books unrelated to training. Ironically, rather than curtailing his performance on the ice, these other activities thrust him forward. "Creating meaning and value in life outside of the speed skating oval helped me get through tough training periods," he writes. "When the training wasn't going great, perhaps something else in life did and that cheered me up." Later on, when van der Poel became more successful, garnering increasing media attention, the other parts of his life kept him grounded. "I knew who I was and I was not just a speed skater," he explains.

Perhaps the greatest advantage of van der Poel's fluid identity is that he became less fragile to the inevitable ups and downs of his career. He writes that diversifying the sources of meaning in his life helped him "to face the horrific fact that only one athlete will win the competition and all the others will lose; that injury or sickness can sabotage four years of work." Paradoxically, it was only when van der Poel became comfortable with the idea of change and disorder that his skating became more relaxed, stable, and fun. One day van der Poel was an Olympian training for seven hours. Another day he was a regular guy with normal friends and normal hobbies. Whatever physical fitness he may have lost by compromising some specificity in his training and recovery, he gained tenfold in mental fitness from his newfound freedom and ease.

Describing the positive impact of his expanded sense of self, van der Poel writes, "There was no longer anything to fear."

Unlike other types of matter, fluid contains both mass and volume but not shape. This allows it to flow over and around obstacles, changing form while retaining substance, neither getting stuck nor fracturing when unforeseen impediments manifest on its path. Cultivating a *fluid sense of self* allowed van der Poel to do the same. By developing and nurturing other parts of his identity, he could flow around bad training days and defeats, over media hype, and through illness, injury, and fatigue.

Van der Poel's fluid sense of self protected him from the mental health struggles so many Olympians face, particularly when the whole of their identity becomes entangled with their sport. A large body of research shows that when there is too great a fusion between one's identity and their pursuit, then anxiety, depression, and burnout frequently result. This is especially true during periods of change and transition, when one's dominant sense of identity feels at risk. World-class sport may be an extreme example, but it is a pattern that holds true in all lines of work and all walks of life: if you want to be excellent and experience something fully, then you've got to go all in, but only to a point. If your identity becomes too enmeshed in any one concept or endeavor—be it your age, how you look in the mirror, a relationship, or your career—then you are likely to face significant distress when things change, which, for better or worse, they always do.

None of the above is permission to be laissez-faire or go through the motions. Van der Poel certainly didn't. He trained *hard* and became the best in the world. Caring deeply about the people, activities, and projects you love is key to a rich and meaningful existence. The problem is not caring deeply; it is when your identity becomes too rigidly attached to any single object or endeavor. You want it to be attached enough, but not too much—a crucial concept that is simple to understand yet hard to practice, which is why we'll focus on it in the proceeding pages.

But first, some notes on terminology. Throughout this chapter, we are going to use the words *self, ego,* and *identity* interchangeably. For our purposes, *ego* will be defined according to its original use in psychology, not *look how great I am* but simply as *I am*. When the ego contracts around a single object, it tends to hold on tight and it hates to die. Fortunately, a fluid sense of self doesn't require the ego's demise, just that it loosen its grip and widen its reach. As with actual fluid, which depends upon the bonding of different atoms, a fluid sense of self, one that is both rugged and flexible, depends upon the successful bonding of our own unique parts, a concept that we'll turn to next.

Persistence Favors Complexity

Perhaps nothing represents a continuous and unrelenting cycle of order, disorder, and reorder on a grander scale than evolution. For long periods of time, Earth is relatively stable. Sweeping changes—warming, cooling, or an asteroid falling from space, for example—occur. These inflection points are followed by periods of disruption and chaos. Eventually, Earth, and everything on it, regains stability, but that stability is somewhere new.

During this cycle, some species get selected out. Others survive and thrive. Species in the latter group tend to have high degrees of what evolutionary biologists call "complexity."

Complexity is comprised of two elements: differentiation and integration. Differentiation is the degree to which a species is composed of parts that are distinct in structure or function from one another. Integration is the degree to which those distinct parts communicate and enhance each other's goals to create a cohesive whole. Consider *Homo sapiens* (you and me), by far the most abundant and widespread species of primate. We have large frames, four limbs, opposable thumbs, body temperature that is somewhat resistant to external conditions, good vision and hearing, digestive tracts that can accommodate a variety of nutrients, and the capacity for language and understanding. In other words, we are a highly differentiated species. But we also have enormous brains and advanced nervous systems that integrate all of these parts into a cohesive whole. The combination of these qualities—widespread differentiation and strong integration—makes us a decidedly complex species. Our complexity is how we got here today and why, hopefully, we'll stick around for at least a bit longer.

Though change at the individual level, the primary concern of this book, is different than change on an evolutionary scale, there is still much we can learn from evolution's foundational principles, lessons that apply to the horizons of our own lives. If we want to survive and thrive during ongoing cycles of change and disorder, then we, too, can benefit from developing our own versions of complexity.

Ginger Feimster was struggling. Though she had grown up wealthy in Belmont, North Carolina, her family had lost nearly everything to gambling. There were times when the electricity would be cut off and it was hard for her to put food on the table for her three children. She wasn't getting along with her husband, Mike, and they divorced not long after their third child, Fortune, was born. To help her through the tough times, Ginger relied heavily on her strong Christian faith, which had always been central to her identity. After the divorce, she doubled down and began dating especially religious men, growing more invested in the church, her beliefs becoming increasingly fervent as a result.

Despite her meager financial position, Ginger was adamant that her only daughter, Fortune, grow up like a proper Southern lady, just as she had. With whatever limited funds she saved, Ginger enrolled Fortune in finishing school, a relic of Southern culture in which girls are taught social graces, etiquette, and upper-class cultural rites, such as "preparation for entry into society." Ginger found motivation and purpose in her belief that Fortune would make her proud when she debuted as a lady at the Gastonia debutante ball, which Fortune did in 1998, at the age of eighteen. Ginger was thrilled with the woman her daughter was becoming. It gave her life meaning and made all her sacrifices seem worth it. Fortune, on the other hand, wasn't so sure. Her main motivation in going through these motions was pleasing her mother, whom she loved dearly.

A few months after the debutante ball, Fortune headed off to Peace College, a small school affiliated with the Presbyterian church in Raleigh, North Carolina. She excelled there, playing soccer and tennis, serving as student body president, graduating summa cum laude, and speaking at her class's commencement in 2002. All of her extracurricular hours were filled with activities, she told herself, so there wasn't much time to date boys. Shortly after college, Fortune moved to Los Angeles to pursue a career in show business. It was there that she watched a movie on Lifetime,

The Truth About Jane, which featured a lesbian protagonist. Fortune was captivated, and in a single moment she realized what had been building up inside of her for years: she, too, was gay.

Understandably, she was nervous to share this news with her mother. "Growing up in the South, we loved two things, and that was church and Chili's. Everyone in the South goes to church; there's one on every corner," recalls Fortune, remembering the community in which she grew up during the early aughts, long before homosexuality was broadly accepted anywhere, let alone in the South. "I decided to take her to my favorite Chinese restaurant to break the news because, I thought, 'Well, even if she disowns me, I can at least eat some crab rangoons.'"

Initially, after hearing that her daughter was gay, Ginger became stone faced. "I was like, 'Oh, man, does she hate me now?'" remembers Fortune. There was a long moment of silence before Ginger's stone face finally gave way to a big smile: "We are going to Hooters," she said. It was Ginger's way of telling her daughter that she accepted and loved her no matter what.

Reflecting on that moment in a podcast interview, Ginger says, "Of course when you have a daughter, you imagine her in a white wedding dress with her bridesmaids, being married in a church, and it's just a beautiful vision for a mother to have. So when Fortune turned out not to be straight, I knew that vision was not going to be a reality. None of it was disappointing; it was just different. And I don't see how you could be disappointed in anything that your child does . . . I thought, 'Okay, that's fine.'"

Fortune went on to become a popular comedian with her own Netflix special, *Sweet and Salty*. Her acts focus on her Southern roots and sexuality. In 2020, she married her wife, Jacquelyn. Ginger became a staunch LGBTQ advocate, all the while remaining a devout Christian and Southern lady. In a situation where many cling to a rigid sense of self, far too often leaving broken families in their wake, Ginger became fluid. She integrated the different parts of her identity—Christian, debutante, Southerner, parent of publicly gay child, LGBTQ activist—under the banner of being a loving mom. When her life was disrupted and changed in ways that she never could have imagined, Ginger became more complex. As a result, her life is far more textured and meaningful.[*]

The speed skater Nils van der Poel realized that he'd face all manner of evolutionary shocks in his own life, from winning big races, to losing big races, to getting injured, to aging out of top fitness, to retiring from competitive racing at an age when most people in traditional careers still have their peaks long in front of them. Van der Poel's uncanny wisdom wasn't just in apprehending this, but in doing something about it. "It was a challenge for me to discover that without my sport I did not have many friends," he writes. "Today I'm very happy for all the friends I've made on all of those rest days . . . They shed light upon my life from a new perspective . . . I believe that it was the value I created outside of the sport, and not the success within it, that made it worthwhile to live in this manner . . . In the long term, the meaning I created apart from my sport made me like my sport more because it suddenly enriched my life rather than limiting it." By reaching beyond the confines of the speed skating oval, van der

---

[*] Only after much deliberation did I decide to include this story. In many parts of the world having a gay family member can still feel in conflict with someone's values. Hopefully in the near future I won't need to include stories like this because they'll be no big deal. Gay is gay; straight is straight; and it doesn't impact someone's being other than who they want to sleep with—which is precisely what Ginger so quickly realized about her daughter.

Poel differentiated his sense of self. By finding a way to have his life outside the sport support his life inside the sport, and vice versa, he integrated it.

After dominating at the 2022 Beijing Winter Olympics, winning gold medals in both long-distance races and setting a world record, van der Poel suffered neither grandiosity nor post-Games depression. Instead, he published his sixty-two-page training manifesto and hung out with his friends. But there's even more to his story of complexity. Having learned details about the Chinese government's restrictions on free speech, tempering of dissent, and treatment of ethnic minorities, he decided to take an act of protest. At a small ceremony in Cambridge, England, less than a week after the Games concluded, van der Poel gave his gold medal to Angela Gui, the daughter of Gui Minhai, a Chinese-born Swedish publisher who is serving a ten-year prison sentence in China for distributing books that are critical of Beijing. "I just hope human rights get to stand at the center of this," van der Poel said of the decision to present his gold medal to Angela. "It's surreal giving away what you fought for your entire life, but it also brings a lot more value to the journey—that it's not just me skating around in circles."

It is hard to imagine a moment when someone's identity is at greater risk for becoming narrow and rigid than setting a world record and winning two gold medals. The fact that van der Poel literally shed that, too, is a perfect metaphor for how he's approached his path. At moments when his identity as "world-class speed skater" is at risk of becoming too entrenched, he actively differentiates and integrates his sense of self. His complexity has helped him to navigate the significant changes he's already faced, and it will undoubtedly bolster him for those yet to come. As van der Poel quoted from Carl Jung at the beginning of his training manifesto, "It seems that all true things must change and only that which changes remains true." He seems to be taking Jung's advice to heart; within it lies wisdom for all of us.

Independence Versus Interdependence

In the middle of the twentieth century, the psychologist Kurt Lewin developed what he called *field theory*. In short, field theory says that all behavior is a function of a person and their environment: people have dynamic thoughts, feelings, and impulses that emerge from the interaction between their brains, their bodies, and their surroundings. Academic papers on field theory are some of the most cited in all of psychology. It is hard to express how profound this view was when Lewin first developed it, at a time when the distinct and separate individual was at the center of psychology. And yet, intuitively, field theory makes immediate sense. You are a very different person when you are with your friends, at work, on vacation, staying at your mother-in-law's house, listening to beautiful music, caught in a downpour, on a sunny beach, scrolling social media, and so on. Few people would argue with this, but when it comes to how we conceive of our "selves," hardly anyone, at least not in the West, considers the role of their environment, let alone weighs it heavily. Rather, when asked to define their "selves," the vast majority of people respond narrowly within the confines of their own skin and skull. When people ask what Enneagram number or Myers-Briggs personality type you are, the most accurate answer is probably some version of "it depends": on where you are, who you are with, whether or not you are hungry, how well you slept the previous night, whether you exercised that morning, and a variety of other factors.

The work of Hazel Rose Markus and Alana Conner, both behavioral scientists at Stanford University, explores cross-cultural differences regarding a range of topics. On the matter of identity, they have found that, broadly speaking, people in the West favor an *in*dependent interpretation of the self and people in the East favor an *inter*dependent interpretation of the self. "Independent selves view themselves as individual, unique, influencing others and their environments, free from constraints, and equal (yet great!)," write Markus and Conner in their book, *Clash!: How to Thrive in a Multicultural World.* "Interdependent selves, by contrast, view themselves as relational, similar to others, adjusting to their situations, and rooted in traditions and obligations."

Consider a study designed by Mutsumi Imai from Keio University and Dedre Gentner from Northwestern University. Participants come into a lab and are presented with a pile of sand shaped like the letter *S*. Next, the participants are shown two more situations: one contains a plain pile of sand in no specific shape; the other contains fragments of broken glass arranged to look like an *S*. Finally, the participants are asked which of the two second situations more closely resembles the first. Imai and Gentner have presented this dilemma to thousands of people across different cultures. They've repeatedly found that Western participants are far more likely to choose the glass, and Eastern participants are more likely to choose the pile of sand. Put differently, Western participants initially see an *S* (an object) that happens to be made out of sand (the field), whereas Eastern participants initially see sand (the field) that happens to be in the shape of an *S* (the object).

It is important to note that neither view is inherently better or worse. "Across [our] many studies, we find that independent and interdependent selves are equally thoughtful, emotional, and active, but often have subtly different thoughts, feelings, and actions in response to the same situations," write

Markus and Conner in *Clash!*. What matters is that different people perceive the exact same situations in different ways, based on the lens they are looking through.

That there are such stark and predictable cultural differences suggests that both of these lenses, independent and interdependent, are largely, if not wholly, learned. Nobody is born seeing the world in a specific way. We adopt our perspective over time. With awareness that multiple lenses exist, we can begin to see the world in multiple ways. We can ask ourselves what lens we are looking through and whether or not it is the best prescription for a given situation.

It is another example of non-dual thinking: the most fluid, and I'd argue the most advantageous, way to conceive of one's "self" is that it can be both independent *and* interdependent. Though these two types of self are often thought of as exclusive, they are most powerful when carried together, like two different tools in a tool kit. In some circumstances it can be helpful to embody an independent self that is unique, influencing, and highly autonomous—for instance, if you are working mostly alone on a big project where the environment is largely in your control. In other circumstances you probably benefit more from adopting an interdependent self that is relational and adjusting, such as when you are working with others or in an unstable environment with many forces beyond your control.

The unity of content and context, of individual and environment, is a central topic of exploration for Quito-based musician, producer, and DJ Nicola Cruz. Though some have dubbed his mesmerizing sound "Andean Step," his work defies genre and embodies non-dualism. It emerges from the confluence of modern electronic downtempo beats and traditional, ancestral sounds. Cruz's music has captivated an international audience, introducing the world to Ecuador's indigenous rhythms and folklore. When NPR reporter Sophia Alvarez Boyd asked Cruz what inspired him to integrate South American storytelling and mythology into his music, he responded, "Living in a place like Ecuador, it just feels natural. All around folklore and roots are quite present. You turn on the radio, and you listen to folkloric music." He echoed the sentiment in another interview, stating, "Ecuador is simply a folkloric country."

More than informing his creative process, Cruz's context *is* his creative process. He selects recording locations with the intention that they shape his work. From a warehouse in New York City, where he recorded his track "Colibria," to a cave in the Ilaló volcano, where he recorded "Arka," external settings factor widely into the production of his music. "Working with different music environments is one of those things that gets me very inspired," Cruz told *Rolling Stone*. "We get interesting reverb, surprises that make it into the recordings. That's what I look for in my music." You cannot separate *where* Cruz is from *who* Cruz is. He has absorbed the environments of his home city, Quito, his home country, Ecuador, and the entire continent of South America into his identity as a creator, and his creations necessarily emanate from them. When you listen to Cruz's music you hear an underlying truth: none of us exist in isolation from the environments we inhabit.

The same theme is true on the playing field. In a crucial paper on talent development in sport, the researchers Duarte Araujo and Keith Davis argue that skill acquisition is best characterized as "the refinement of adaptation processes, achieved by perceiving the key properties of the surrounding layout of the performance environment in the scale of an individual's body and action capabilities." The athletics coach Stuart McMillan, who has guided over thirty-five athletes to Olympic and World Championship medals, simplifies it: "Skill is not some *thing* to develop or acquire; rather, it is an emergent interaction with an ever-changing environment," he says. The best athletes find ways of working in concert with their surroundings, adapting themselves and their performances. They, too, are both independent *and* interdependent.

We've just examined the benefits of developing a fluid sense of self in relation to an ebbing and flowing environment, which we are both separate from and part of at the same time. We'll proceed by going a layer deeper, delving into a big, deep, important, and intellectually challenging topic: how to think about identity when everything is always changing, including us. Eventually, we'll confront the question of whether or not there is such a thing as an enduring "self" at all.

Before we begin, I want to propose some guardrails for our exploration. A self that is completely interdependent and unbounded with no confines whatsoever may, in a protected setting and for a short period of time, represent a profound spiritual awakening or enlightenment. But outside of that protected setting and limited duration, it tends to look a lot more like chaos or psychosis. Meanwhile, a self that is completely independent—that thinks it is largely unchanging, separate from, and in control of everything around it—may also be beneficial in very specific conditions, such as trying to swim across an indoor pool as fast as possible. But beyond that, conceiving of oneself in this way is liable to result in crippling neuroticism, anxiety, loneliness, and eventually, depression. What we are going to explore in the upcoming sections is the middle ground: how to conceive of yourself as something that is both singular and stable on the one hand *and* porous and constantly changing on the other. As you'll soon see, there are immense benefits to adopting this view.

A Strong, Stable Identity Through Change

One morning while I was in the initial stages of writing this book, I was at the gym testing my strength across the three big movements for which I train: squat, bench press, and dead lift. I'd been training somewhat seriously for the past eighteen months, and it was my first chance to really go for it and attempt to set a new mark. As I approached the barbell for my first lift, I struggled to find an extra gear, one that in the past I could easily access. It sounds bad to say, but whether or not I made the lift just didn't matter as much as it used to.

Rewind to seventh grade, when I desperately wanted to play football but my parents would not let me. Then, I got jumped on the side of the road by two high school punks. It was terrifying. I became anxious and scared of being outside alone in my own neighborhood. However, there was a silver lining: my parents decided to let me play football, thinking it might help to bolster my confidence.

I gave my all to football. I was first in and last out of the weight room every day. I got strong from training and felt safer and more secure in my body. Girls started liking me, and as much as I cringe saying it now, it probably had more to do with the size of my arms than anything else. I became captain of the varsity football team and we had the best consecutive two-year record (17–3) in the previous four decades of my high school's history. I got recruited to play football at smaller college programs, and although I ultimately decided to attend the University of Michigan (where I was not talented enough to play), football and strength training remained a huge part of my identity during my formative years. You could easily argue they *were* my identity.

At Michigan, I couldn't go to football games. It felt pointless to be in the stands instead of on the field, too close to something the loss of which I was still grieving. So, I did a total 180 and found myself training for endurance sports, starting with marathons and eventually triathlons. Though it was different from football, I kept my primary identity as an athlete intact. At the end of my junior year, the girl I had been dating since I was a freshman told me she still had feelings for another guy and dumped me. Though looking back this was clearly for the better, at the time it was deeply painful. I threw myself into triathlon training with full force—not so much because I loved swimming, biking, or running, but because it was a good analgesic.

Fast-forward over a decade and I got back into strength training when my wonderful wife, Caitlin (thank goodness the other girl broke up with me), was deep into her pregnancy with our first child. Training for triathlons and marathons was consuming too much of my time and energy and I was getting injured far too frequently. I wanted a physical practice that would better fit into my new life as a father. At first, I trained without much structure. It simply felt good to be back in the gym. A few years later, however—out of the thick of infant parenting and into the thick of the pandemic, when there wasn't much else to do for leisure—I decided to focus more on lifting weights. I set up a modest gym in our garage and trained four to five days a week for about sixty to ninety minutes each session. I was far from an elite athlete, but more advanced than a total newbie.

Let's revisit my strength test, and I've got a theory for what happened. The person who walked up to the barbell to squat in 2022 was very different from the insecure kid in the high school weight room or the narrow adult on the triathlon course. For the first time in my life, my performance in sport was not core to my identity. I am a husband. I am a father. I am a writer. I am a coach. I am a reader. I am a friend. I have a spiritual practice. I love German shepherds. And only then, perhaps tied with, or even slightly below, "I enjoy long walks outdoors" comes strength training. The reason I couldn't find that extra gear, the reason I felt like whether or not I made the lift didn't matter as much as it used to, is precisely because it doesn't matter as much as it used to.

In the past, a test of strength or fitness felt like everything. Performing at my peak was a matter of self-preservation; it was nonnegotiable. Though I (mostly) enjoyed the sports, I was competing in threat mode. All the resources available to protect my identity—i.e., make the lift or win the race—were easily marshaled. At present, however, I am not reliant on sporting success to validate my identity. Whether or not I make a lift has much less bearing on my self-worth.

Although at first I was confused and frustrated—don't get me wrong, I still care about my performance, only in a different way than I once did—after some reflection I found myself curious and even excited about my newfound relationship with sport. I get to experience up close and personal the gradual transformation of my identity from within. Is there a way to be a performance-driven athlete at specific times but not at others? Can I learn how to flip the switch on demand? And beyond that, who am I if not a performance-driven athlete? What does it mean to know that I'll likely go through similar transformations with other parts of my identity in the future?

My athletic prowess is nowhere close to that of someone like the speed skater Nils van der Poel, but the underlying tension I faced is the same. It is not just me and it is not just sports. Whenever I share this story with other people, they nod in agreement. Perhaps it is someone whose identity once focused solely on making art or on their work as an entrepreneur or physician. Or a person who is recently separated or divorced grappling with their former identity as husband, wife, or partner. Other times it is an older adult whose central identity used to be "parent to young kids," but now their kids are grown. In all of these instances, we are the same as we used to be but also unequivocally different. It is a tricky dilemma to grapple with, and yet it is one that just about everyone faces, and often repeatedly, throughout their lives. Fortunately, modern science and ancient wisdom can help us to make sense of it all.

Developing a Rugged and Flexible Ego

Jane Loevinger was a twentieth-century American psychologist who, along with her colleague Erik Erikson, pioneered the study of ego development. The work of Loevinger, who died in 2008, is crucial to understanding how one develops a fluid sense of self and a rugged and flexible identity. Loevinger described the ego as an unfolding process, not a static entity. Within that process, she identified nine key stages, starting during infancy and maturing all the way through adulthood.

Early on, there is hardly any ego to speak of. An infant is completely dependent on everyone and everything around it, most notably their caretakers and home environment. As a child matures, they start to develop a sense of self that is separate, an extremely important milestone that most kids reach around age two. Gradually, that separate sense of self builds confidence, a key precondition to exerting one's will on the world (e.g., feeding oneself and using the potty). Gaining a separate sense of self also helps children become more social by developing what psychologists call "theory of mind," or realizing that the entire world doesn't revolve around them, that other people have their own distinct wants and needs, too. From there, young adults learn about rules and social norms, as well as how to navigate and protect themselves from threats in their environment. (This describes me in high school, getting strong after being jumped on the side of the road.)

As we progress through adulthood, our egos become more refined. If we are lucky and reach the later stages of Loevinger's model, we go from craving external validation and outward achievement to prioritizing internal meaning and fulfillment. In the final stage of Loevinger's model, which few reach according to her research, the ego demonstrates deep empathy as well as self-acceptance. It cherishes its own idiosyncrasies as well as those of others, and it understands both its separateness from *and* connection to everything around it. Some developmental psychologists have proposed an additional stage, which they call "unitive." Here, the ego embraces the fact that it is both solid and pliable. It is able to integrate these two seemingly contradictory states into a cohesive whole.

As is the case with all models, Loevinger's stages of ego development has been subject to criticism surrounding its accuracy and applicability. That said, it has stood the test of time. Loevinger was meticulous in her measurement of each stage, using a validated survey instrument that has proved reliable cross-culturally, from Australia to India.

I find Loevinger's stages of ego development useful for two main reasons. First, they recognize that our sense of self is not static but dynamic, or in Loevinger's words, that "the ego is an unfolding process." Second, each stage of ego development works great until it gets in the way. The arc of her model could be summarized as follows: your survival depends upon developing a distinct and strong sense of self, but as you grow older and gain wisdom, that distinct and strong sense of self can start to become a hindrance, at least in some situations. The same ego that helps us meet our basic needs, healthfully separate from our caretakers, and protect ourselves from threats can also cause feelings of isolation, anxiety, and existential distress. The essential skill, then, is to realize when our ego's current manifestation is of service to us, and learn to leave it behind when it is not. When I'm at an intersection and the light turns from red to green, it is very important that I identify with an ego that is separate and in control, so that I can proceed to hit the gas and get on my way. Same when I'm trying to make a big lift at the gym. When I become an empty nester or I'm sick or on my deathbed, however, I'd much rather identify with an ego that is vast, interconnected, and not an over-controller. These may be extreme examples, but they elucidate a crucial point: even the ego itself can be a fluid and flexible concept, if we choose to make it one.

Conventional and Ultimate Selves

When the historical Buddha was teaching across Asia, he was approached by a wanderer named Vacchagotta, who asked him point blank whether or not there is a self—quite the question. Here is the scholar Bhikkhu Bodhi's translation of the scene, as it is recorded in the Pali Canon, one of the oldest remaining Buddhist texts:

> "Now then, Venerable Gotama, is there a self?" asks Vacchagotta. When this was said, the Buddha was silent. "Then there is no self?" replies Vacchagotta. A second time, the Buddha was silent. Then Vacchagotta the wanderer got up from his seat and left.

Later on, Ananda, the Buddha's loyal attendant and reliable sidekick, asks him about the situation:

"Ugh, Buddha, that seemed like a pretty important question that Vacchagotta the wanderer just asked you. What was going on back there? Why didn't you answer him?" (Translation mine.)

Since the Buddha's response is quite important, let's go back to Bhikkhu Bodhi for the translation:

"If, Ananda, when I was asked by the wanderer Vacchagotta, 'Is there a self?' I had answered, 'There is a self,' would this have been consistent on my part with the arising of the knowledge that 'all phenomena are nonself'?" says the Buddha.

"No, venerable sir," says Ananda.

The Buddha continues, "And if, when I was asked by him, 'Is there no self?' I had answered, 'There is no self,' the wanderer Vacchagotta, already confused, would have fallen into even greater confusion, thinking, 'It seems that the self I formerly had does not exist now.'"

The Buddha's silence means he saw no useful answer to the question.

The Buddha's encounter with Vacchagotta and his subsequent discourse with Ananda has become one of the most discussed passages across Buddhist texts. Though various modern scholars and schools offer slightly different perspectives, the most common one, and I'd argue the most useful, goes something like this: There is a "conventional" self, which is the self that is reading or listening to this book right now, the self that takes control and drives through intersections. The conventional self is entirely real and important. Without it, we could not navigate day-to-day life. But there is also an "ultimate" self, which is the self that is connected with everyone and everything around it, including the food it eats, its prior experiences, the genetics of its ancestors, the air it breathes, the children it raises, and so on. The ultimate self is every bit as true as the conventional self. Both can, and do, exist at the same time. It is an entirely rational and empirical argument. It only makes people's heads explode because we are so accustomed to this-*or*-that thinking when so many profound truths require this-*and*-that thinking.

The Buddha's enlightened, deeply true, non-dual self is strikingly similar to Jane Loevinger's highest stage of ego development: the ego that understands its fluidity, that knows when to leave itself behind without losing its sense of self altogether.

We run into trouble not when we have a strong identity but when that strong identity becomes too rigidly attached to any one pursuit, person, or concept, including how it views itself. Thus, it is advantageous to hold our identities in two ways at the same time. There is the conventional self that is distinct, stable, and here right now. And there is the ultimate self that is constantly changing, that transcends any one endeavor. Keeping the second self in mind frees us up to more fully excel with the first, since we become less apprehensive about acute failure and change. This is precisely what happened to the speed skater Nils van der Poel. As he developed a more fluid identity, he enjoyed competing more; remember, in his own words, he had "nothing to fear." Ginger Feimster's fluid sense of self allowed her to transcend her historical story about what it meant to be a devout Christian and Southern mom. And I am in the process of discovering a new relationship with performance-driven sport, and thus a new relationship with myself.

We Contain Multitudes

Terry Crews grew up in Flint, Michigan. Throughout his childhood he demonstrated a penchant for art. By age eight, he was excelling on the canvas and also at playing the flute. This continued through middle school, and he received a scholarship to the prestigious Interlochen Center for the Arts boarding school in northwest Michigan. The only thing holding Crews back from attending was that he was also pretty good at football. Actually, that's an understatement. He was dominant. And so he stayed at his conventional high school and found success on the gridiron.

Crews ended up attending Western Michigan University, where he was awarded two scholarships, one in art and the other in athletics. He had a stellar college football career and was drafted to the NFL's Los Angeles Rams in 1991. When he wasn't slaying opponents as a linebacker, he maintained his artistic sensibilities by drawing sketches of his teammates. In 1997, after seven arduous seasons, Crews retired from football. At that point, he had been playing for the Philadelphia Eagles, but he and his wife decided to move back to Los Angeles, where Crews hoped to pursue an acting career. Unfortunately for him, executives in show business could not see how his background in football would be an asset. After a year of rejections, Crews took jobs sweeping a factory floor and working security at nightclubs.

As is the case for so many athletes, Crews' transition out of sport was challenging. He went from being a standout on the football field to being just another guy. "[You realize] you are not who you think you are. Because, you are known as an athlete, you are known as this and you are known as that, and then all of a sudden, you have to rebuild your life... It's very strange, it's very foreign," he explains.

In 1999, through his connections at the nightclub, he got word of an audition for a new show, *Battle Dome*. The show was a bit like wrestling, and his pro-athlete physique made him perfect for the role. After a long and drawn-out audition process, Crews got the part. Following *Battle Dome*, he continued auditioning, taking small parts in movies, and making new connections. Slowly but surely, his acting career gained traction, and he won roles in films and shows such as *White Chicks, The Longest Yard, Everybody Hates Chris, Idiocracy, America's Got Talent,* and *Brooklyn Nine-Nine*. He attributes the persistence that got him to where he is today in Hollywood to the discipline he learned as a football player.

"It's kind of weird because I look at entertainment, and my football career, the ups and downs, the ins and outs to how hard it was, it really prepared me for entertainment in that I could take rejection, I could go to an audition and realize that it wasn't about me," says Crews. "I have to say, being in the NFL for seven years got me ready for Hollywood. It did. You have to learn how to take a punch."

In his book *Range,* the science writer David Epstein makes a compelling case for the benefits of being a generalist. Whereas a specialist focuses very narrowly on a particular subject matter, a generalist seeks a broad array of diverse experiences. Epstein cites hundreds of studies pointing toward the advantages of the latter, ranging from increased creativity, to better health and fitness, to enhanced problem-solving skills. Whether you want to be a scientist, athlete, artist, writer, entrepreneur, or businessperson, the evidence is clear: it's useful to be a generalist, or at the very least, to go broad before you go narrow. Play multiple sports growing up and you are more likely to make it to the pros as an adult. Try different styles of art and you're more likely to create a masterpiece. Study diverse topics and you're more likely to stumble upon a scientific breakthrough or new way of solving a business or management problem.

Range is a wonderful book, one of my favorites of the past decade. I suspect part of the reason it has been so well received is because we understand its message intuitively, even if modern society is constantly telling us the opposite. Narrow specialization works fine in the short run but is neither a good nor healthy strategy in the long run. Like Terry Crews, it is better to conceive of yourself fluidly, to develop an identity that has range. Since *Range* was published in 2019, additional studies have shown that excellence in a specific pursuit often follows a sampling period in many others. In the parlance of behavioral scientists, it is beneficial to first "explore" many aspects of your identity and skills before you go on to "exploit" any particular one. What's more is that we can repeat this cycle throughout our lives. There is a reason that perhaps the most well-known stanza of American poetry is the following, from Walt Whitman:

> Do I contradict myself?
> Very well then I contradict myself,
> (I am large, I contain multitudes.)

In addition to the external benefits of being a generalist, there's an enormous internal one. You become increasingly rugged and flexible. If you can learn to define yourself broadly, then change—be it aging or retirement, gain or loss, success or failure—becomes less threatening. You can take a hit in one part of your identity without losing others. In the next chapter, we'll take this fluid sense of self—which is large and contains multitudes, independent and dependent, differentiated and integrated—and learn about the importance of developing rugged and flexible boundaries to guide its unfolding path.

Cultivate a Fluid Sense of Self

- Like water, a fluid sense of self can go into and fill any one space; but it can also flow out of that space when necessary, changing shape without changing form.
- A fluid sense of self is non-dual; it is:
    → not differentiated *or* integrated, but differentiated *and* integrated
    → not independent *or* interdependent, but independent *and* interdependent
    → not separate *or* connected, but separate *and* connected
    → not conventional *or* ultimate, but conventional *and* ultimate
- The more we can conceptualize our identities non-dually, holding all of these contradictions at once, the better off we'll be and do.
- By conceiving of ourselves in a fluid manner, change, be it internal or external, becomes less threatening; our identities become more rugged and flexible and thus better able to endure and persist over the long haul, including throughout countless cycles of order, disorder, and reorder.

4

Develop Rugged and Flexible Boundaries

Picture in your mind's eye a river. It is a concrete and observable phenomenon. Yet it is also always flowing. An essential part of a river is its bank, which serves as a container to hold the flow and to provide it with direction. Without a bank there would be no river. You'd merely have random water instead. It can be helpful to think about our identities in the same way. The flow represents our fluidity, that we are constantly changing, moving this way and that. The bank represents our rugged and flexible boundaries, which hold and organize the flow, creating a distinct and observable path. In the previous chapter, we discussed how to cultivate a fluid sense of self. In doing so, we focused mainly on the flow. Now, we'll examine the banks, learning how to define and apply the edges that hold our identities together and give them shape over time.

Someone who illustrates this concept well is Georgia Durante: model, turned mafia driver, turned stunt actress, turned entrepreneur and author. During her teenage years in the late 1960s, Durante was featured in advertisements for Kodak cameras. She was also a frequenter of Sundowners, a mafia-run nightclub in New York City, which injected excitement and intensity into her otherwise girl-next-door life. Even so, her evenings at the club tended to be uneventful. That all changed one fateful night, when a man was shot in front of her. "I'm there and five feet away from me, this guy pulls out a gun and shoots the guy next to him. . . . Everybody scattered, and the guy fell to the floor," Durante recalled in an interview with NPR.

Within seconds, the owner of Sundowners threw Durante a set of keys and told her to bring the car around. "Georgie-Girl, go get the car, bring it up," he shouted. The owner, his entourage, and the injured man got in, and she sped them to the hospital, arriving in record time. After they dropped the wounded man off, the mafia members in the car kept talking about how impressed they were with her driving. Following some low-toned mumbling, they offered her the chance to do "driving work." It started with Durante picking up and delivering packages, but as the mafia further observed her extraordinary driving skills, they began sending her more dangerous work. Eventually, Durante became a getaway driver for robberies and other significant crimes. The money was good, and she enjoyed the high-speed lifestyle. If Durante embodied any one attribute, it was intensity.

Some years later, though, a mob war broke out, and Durante knew she'd have to skip town. By then, she had a seven-year-old daughter and was married to a mobster who was becoming abusive. The situation in New York quickly became untenable. They fled to San Diego, California. There, her husband's abuse grew more fervent until one day, Durante worked up the courage to leave. With little more than seven dollars in her wallet, she and her daughter drove to Los Angeles. They lived in her car, stealing food from convenience stores merely to survive. Eventually, Durante and her daughter moved in with an old friend in Brentwood. It was imperative that they lay low, lest they be discovered by the mob or her abusive husband. But Durante also had to figure out how to make money. She was in a jam, and for the first time it was one that she couldn't drive her way out of—or so she thought.

One afternoon, while passing time watching television on her friend's couch, she realized there were tons of car commercials that included devastating curves and winding cliff roads. During these commercials, you rarely, if ever, saw the driver. And then it clicked. The perfect fit. An anonymous job in which she could apply her fierce intensity and well-honed skills. Relying on her mafia networking savvy, Durante gained intel on where the shoots were taking place, and she began showing up on location, pleading with directors to give her a shot at driving. At first, they all responded with hard nos, writing her off because they didn't believe a woman could cut it as a stunt driver. Nevertheless, she persisted, and a director finally gave her a chance. She proceeded to sweep him off his feet.

Durante became known as an elite driver in Hollywood, and she began getting hired for an increasing number of jobs. It wasn't long before she was doubling as Cindy Crawford in Pepsi commercials. The demand for her expertise escalated to the point that she had to turn down work. Eventually, she started her own company, Performance Two, providing stunt drivers for Hollywood productions and nearly all of the major automobile manufacturers. "Life is what it is," Durante writes in her memoir, *The Company She Keeps*. "How we deal with it is what matters."

Durante's ruggedness lies in her intensity and her affinity for driving. Her flexibility lies in how she applied that intensity and the situations in which she drove. If you rigidly dig in, hoping to stay the same and trying to insulate yourself from change, then you run the risk of falling apart. But if you are fluid without any boundaries or direction, then it can become quite confusing as to who you even are. The remainder of this chapter is about guiding the evolution of your identity, or at the very least, setting the general course for your path. It's about confronting and adapting to change and disorder without being so transformed by it that you no longer recognize yourself.

Rugged Boundaries

For all the things in life that you cannot control, there is at least one that you can: your core values, which represent your fundamental beliefs and guiding principles. They are the attributes and qualities that matter to you most. A few examples include authenticity, presence, health, community, spirituality, relationships, intellect, creativity, responsibility, and trustworthiness.

In my coaching practice, I have nearly all of my clients come up with three to five core values. During periods of relative stability, core values act as an internal dashboard, a way to make tangible the characteristics that support you in feeling and doing your best. For each core value, we craft a single sentence to customize it and make it concrete. For example, someone may have a core value of "presence" and define it as *being fully there for the people and pursuits that matter to them most.* The next step is coming up with specific examples for how to practice each core value in daily life. Continuing with "presence," someone might say, *Schedule and complete at least three blocks of deep-focus work on high-priority projects per week,* or *have my partner hide my phone every evening at 7 p.m. and not give it back to me until 7 a.m. the following day so I can be undistracted with my family.*

Core values also play an important role during periods of change, disorder, and uncertainty. When you feel the ground shifting underneath you, when you don't know your next move, you can ask yourself, *How might I move in the direction of my core values?* Or, if that isn't possible, you might consider, *How might I protect them?* For example, if you have a core value of "creativity," you can change jobs, and even change mediums, while still exemplifying your value. I've practiced creativity drafting PowerPoint slide decks for a consulting firm, coaching physicians, hosting a podcast, writing books like this one, and parenting a young child. In Georgia Durante's case, her core value of intensity stayed with her all along, even as her life changed dramatically.

The portability of core values means that you can practice them in nearly all circumstances. Thus, they become a source of stability throughout change, forging the rugged boundaries in which your fluid sense of self can flow and evolve. Nothing can take your values away from you. They provide a rudder to steer you into the unknown, guiding how you differentiate and integrate over time. Yes, core values can, and sometimes do, change; but even then, it is prioritizing and acting on your previous core values that lead you to your new ones. Fortified by your core values, change, disorder, and uncertainty become a bit less threatening and intimidating.

Consider a recent study published in the *Proceedings of the National Academy of Sciences*, in which Emily Falk, a psychologist from the University of Pennsylvania, and her colleagues used fMRI technology to examine what happens inside of people's brains when they are presented with changes that commonly feel threatening. For example, someone who smokes or abuses alcohol may be told they'll need to quit starting tomorrow. Or someone who has never exercised before may be told they'll need to start a workout plan later that day. Individuals who were instructed to reflect deeply on their core values prior to being presented with these scenarios showed heightened neural activity in a part of the brain (the ventromedial prefrontal cortex, or VMPFC for short) associated with "positive valuation," or viewing threats as just manageable challenges. Instead of shutting down or resisting a potentially hard change, their brains were moving them toward engaging with it. Those who were not told to reflect on their core values showed no increased neural activity in their VMPFC. These effects were not confined solely to the lab. The individuals who reflected on their core values went on to successfully work through big changes in the real world at a much greater rate than the control group (in this study, relating to health behaviors, like the examples above).

Let's shift from biology to psychology. At the crux of anxiety generally lies overwhelming worry about change and uncertainty. It is unsurprising, then, that core values feature prominently in a gold-standard treatment for anxiety disorders: acceptance and commitment therapy, or ACT for short. When I spoke with Steven Hayes, the University of Nevada clinical psychology professor who developed ACT, he explained that the main effect of core values lies in the strength and ruggedness they provide. Everything around you—and in the case of anxiety, everything within you, too—can feel out of control; and yet you can still show up and act in alignment with your values.

Anxiety wants you to avoid change and uncertainty, and thus it almost always has a limiting effect on someone's life. But if you can know and trust in your core values, essentially knowing and trusting in the deepest parts of yourself, then you can courageously walk forward into the unknown. Regardless of what you are faced with or what you are feeling, you can lean on your core values for support and to guide your next steps. Hayes and his colleagues have demonstrated the positive impact of core values in hundreds of studies. His psychological findings on the power of core values align almost perfectly with Falk's biological ones. Her fascinating fMRI study merely lets us peer under the hood. Even amidst fear, chaos, and threat, our core values offer ruggedness, strength, and stability.

This subject matter is particularly important, so it's worth a quick summary: Core values are your guiding principles. It is good to have three to five (an extended list of example core values is in the appendix on page 202). Define each of your core values in specific terms and come up with a few ways that you can practice them in day-to-day life. The goal is to take what can seem like lofty qualities and attributes and make them as tangible as possible. When you are confronted with change and disorder, use your core values to navigate into the unknown. Ask yourself how you'd move in the direction of your core values and in what new ways you might practice them. If an outside force is requiring that you leave your core values behind—that there is genuinely no constructive way you can apply them in the new reality—then that is generally a good sign to consider fighting back. Though it is not necessary, it is normal for your core values to change over time. Navigating the world using your current core values is what guides you to your new ones.

Your core values hold your differentiated and integrated, independent and interdependent, and conventional and ultimate selves together, creating a coherent, complex, and enduring whole. They help you to make tough decisions, serving as the boundaries within which you evolve and grow over time, which is the topic we'll turn to next.

Flexible Application

In his decades of theorizing and researching allostasis, Peter Sterling consistently observed the same pattern: organisms that remain healthy and resilient over long periods of time are those that are able to adapt to their changing environments. But they don't do it at random. Rather, their adaptation is guided by their central features and needs. In purely biological terms, this means that most organisms adapt in ways that draw upon their strengths to give them a better chance at eating, drinking, and reproducing. Since the scope of this book extends beyond basic biology and survival, we can take Sterling's findings and apply them more broadly. Our health, longevity, and excellence—our ability to fully manifest our gifts, to feel good and do good—is contingent upon our ability to adapt in ways that protect, and ideally promote, our "central features," or our terms, our core values. However, this does not mean that we always practice them in the same ways. Here, flexibility is essential.

Roger Federer is one of the greatest male tennis players of all time. His career is exceptional for many reasons (he has won 103 singles titles including 20 major championships) but probably most of all for its length. Whereas many tennis players peak in their late twenties, Federer continued to be a dominant force well into his thirties. Yet his trajectory wasn't straight upward. Between 2013 and 2016, Federer, then in his early thirties, suffered a slew of injuries, particularly affecting his back. He did not win a single major championship during that stretch, and he was forced to drop out of tournaments that in the past he would have won easily. Many believed that Federer's age was finally catching up with him.

But then something remarkable happened. In 2017, the year in which he turned thirty-six, Federer had an incredible season, one of the best in his entire career. He won fifty-four matches and lost only five, the best win percentage he had posted since 2006, when he was twenty-five years old. He captured two major championships and was ranked number two in the world. This comeback and Federer's remarkable longevity are attributable to two major factors: first, his unwavering love of tennis and his dedication to competition and excellence; second, his ability to adapt over time. Most people reach a point in their careers when they become resistant to change. They are happy to keep doing things the way they always have. Federer is not like most people.

When he hit his age-related rough patch between 2013 and 2016, Federer made a number of significant changes. He trained and competed less so he could focus more on rest and recovery between big events. He played at the net frequently, which shortened the duration of points so he wouldn't have to run the baseline for hours on end against competitors who were much younger. He learned a one-handed backhand with massive topspin that conserved energy and gave his rival, Rafael Nadal, all sorts of trouble. And he embraced new technology, leaving behind the racket he had played with for most of his career, the one that had made him perhaps the greatest male tennis player of all time, in favor of a new racket with enhanced design technology that all the younger players were already using.

Federer held on to his core values of competition and excellence, as well as his love of tennis. But when confronted with the inevitability of aging, he was flexible in how he practiced them. "You can be stubborn and successful or you can give it up a bit and change things around. For me, it's important to have a bit of both," says Federer. "The back injury in 2013 gave me the opportunity to look at the picture on a broader scale rather than just think I need to get my back straight and then I'll be fine again and we'll go back to status quo... Everything keeps evolving and changing. I've always been quite open about it."

As a result, Federer has had one of the longest and most successful careers of any male tennis player in modern times. He has also become an exemplar for a younger generation of stars, a role that will undoubtedly open new opportunities for him to practice his core values in his retirement, which he announced at the end of 2022, at age forty-one.

In the early 1960s, the American astronomers Arno Penzias and Robert Wilson had the opportunity to get their hands on a giant antenna that had been built by Bell Labs in Holmdel, New Jersey. Originally used to transmit information across long distances on Earth, the antenna was outmoded by new technology in 1962, freeing up the contraption for Wilson and Penzias to use in their research, which was focused on studying radio waves from the Milky Way. The duo were thrilled to have this new instrument in their arsenal.

Shortly after they began working with the antenna, an annoyance put a damper on their enthusiasm. No matter where they pointed it, the antenna transmitted a constant low-level background static that distracted them from the signals they sought. After testing a variety of hypotheses for likely sources of the static—these included the antenna itself, noise from nearby New York City, human nuclear activity, interference from planetary movement, and even pigeons—the two were faced with a reality they could not ignore, at least not if they were going to stay true to the scientific method that they so dearly valued. The low-level static they were picking up was not a bug but rather a feature. It was an important part of the universe that warranted investigation in and of itself.

While other researchers had stumbled upon the ubiquitous hum before, this was the first time the larger scientific community would take it seriously. In the past, many scientists had disregarded the phenomenon because its significance would have only applied to cosmology, the study of the origin and development of the universe. Yet cosmology hadn't been the most well-regarded field. In his book *The First Three Minutes,* the theoretical physicist Steven Weinberg explains that "in the 1950s, the study of the early universe was widely regarded as not the sort of thing to which a respectable scientist would devote their time." But Penzias and Wilson *were* respectable scientists, and their meticulous approach led them to believe that this mysterious hum ought to be taken seriously.

Despite the fact that the noise did not fall under the scope of their original investigations, and even though the noise would have applied to a field they were predisposed to overlook, the duo took their findings to physicists at Princeton University. One such physicist, Robert Dicke, had been working on proving the big bang theory, which at the time was mired in controversy. According to Dicke, the aftershock of the universe's explosive birth would have been perceptible as background microwave radiation. This described precisely the phenomenon that Wilson and Penzias had stumbled upon.

Initially, Wilson wasn't thrilled about the idea that something he'd helped to discover could be used to validate the big bang theory. After all, he subscribed to its rival, the steady state theory, which held that the universe has neither a beginning nor an end. But when paired with earlier scientific findings, Wilson had no choice but to face the music, or more precisely, the low-level background hum: What he'd observed with his giant antenna neither validated his prior beliefs nor helped him to find the Milky Way radio signals he was looking for. What it did do, however, was prove a big part of his worldview wrong.

What Penzias and Wilson initially experienced as an annoyance turned out to be what scientists now call the cosmic microwave background, or CMB. The CMB has not only vaulted the big bang theory into its current position as the dominant explanation for the origin of the universe, but it has also provided all manner of other insights into our universe's history. Today, astronomers use the CMB to determine the total contents of the universe, to understand the origins of galaxies, and to study the very first moments after the big bang.

Wilson and Penzias could have stuck to their initial presumptions and predispositions. They could have decided to ignore the static, writing it off as a mysterious nuisance. They could have also refused to take their findings to a researcher whose cosmological view challenged their own. If they had done this, we might still be deaf to our universe's instructive murmuring. But, the two researchers adhered to their shared core value of the scientific method, choosing to adapt their exploration, and eventually, their conclusions, in alignment with the empirical evidence. The duo's rugged flexibility was rewarded: in 1978, Penzias and Wilson won the Nobel Prize in Physics for their discovery.

The stories of Georgia Durante, Roger Federer, Arno Penzias, and Robert Wilson all demonstrate the power of flexibly applying one's rugged core values.

Ruggedness without flexibility is rigidity; flexibility without ruggedness is instability. Put the two together, however, and you emerge with the supple strength needed to persist and thrive over the long haul—a theme that is true not only for individuals, but for organizations as well.

Population Ecology

In the late 1970s, the organizational psychologists Michael Hannan and John Freeman, then at Stanford University and the University of California, Berkeley, developed a theory called *population ecology*. Population ecology looks at specific industries and examines the births and deaths of organizations over long periods of time. They repeatedly found that when a specific field's operating environment changes, certain organizations get selected out and replaced, either by their current competitors or by new ones that are better suited to the external demands.

To this day, population ecology remains a cornerstone of organizational studies. It is a complex and intricate theory; you could get an entire PhD studying it. But if I had to summarize its three biggest principles, it would go something like this: First, the more rigid an organization's structure, the more likely that organization is to get selected out during periods of disorder. Second, an organization's strength in the short run all too easily becomes its weakness in the long run; if an organization is too calcified around certain attributes or goals, when the environment changes those very attributes or goals often get in the way. Third, the bigger the external change, the more likely it is that all the entrenched organizations in an industry either become extinct or change so much that they are hardly recognizable.

In other words, organizations are like individuals: they struggle to maintain their identities during periods of change and disorder. Some don't change enough. Others change so much that they completely lose sight of what they are. Only organizations that deliberately cultivate their rugged boundaries and then flexibly apply them have a shot at prospering over the long haul.

When Hannan and Freeman developed population ecology in the 1970s, large-scale industrial change was still relatively slow, like tectonic plates gradually shifting. Ever since the proliferation of the internet in the mid-1990s, however, the pace of change in most industries has intensified exponentially. Nearly all of the global economy has become reliant on computers. Moore's Law implies that computing power doubles about every two years. The consequence has been intensified change, more frequent and compressed cycles of order, disorder, and reorder. Skillfully navigating these cycles is critical to an organization's longevity, and the consequences of messing up are significant. It's why Blockbuster Video still appears on PowerPoint slides in boardrooms across the world, representing a horror story that every company now tries to avoid.

Perhaps no industry has been more affected by rapid technological change than newspaper publishing. It used to be that a reader's only option was a print newspaper and an advertiser's only option was a print ad. As a result, in the mid-1990s there were a handful of thriving national and regional newspapers as well as countless local ones. That's obviously no longer the case. Today, newspapers compete with websites, podcasts, social media, videos, and numerous other digital media for advertisers and readers, now called *end users* or *eyeballs*. According to the Pew Research Center, newspaper circulation has declined by 60 percent since 1995, down from around sixty-two million daily copies back then to twenty-five million today. Despite some gains in digital readership, overall newspaper revenues have declined about 66 percent since 2000. And since 2004, employment in the newspaper sector is down more than 50 percent. Whereas the vast majority of newspapers have struggled merely to survive, there is at least one that has thrived.

Newspapers can be polarizing. I'm asking you to put that aside for minute as we delve into a brief case study. Because regardless of your literary taste, politics, or cultural proclivities, it is incontestable that, *as a business,* the *New York Times* has performed exceedingly well during a period of massive disorder and disruption in its industry. In the year 2000, the *New York Times* had about 1.2 million paid subscribers receiving the newspaper, which at the time was predominantly distributed in print. By 2022, the *Times* had over 10 million paying subscribers, the vast majority of whom accessed the paper digitally. But to say accessed the "paper" is a misnomer. The *Times* also draws millions of people to its popular podcasts, crossword puzzles, and cooking apps. Though the *Times* hasn't been immune from declining advertising revenue, the company has remained highly profitable. In 2021, the *Times* reported a net income of 220 million dollars, and at the end of that year, the company's share price reached its highest mark on record at over fifty-four dollars per share, up more than 20 percent from the year 2000.

The *Times*'s remarkable performance against all odds owes itself to the organization's flexible application of its core values. According to the company's website, the *Times*'s values include independence, integrity, curiosity, seeking out different perspectives, and excellence. The values did not change when the organization transitioned out of the twentieth century and into the twenty-first. What did change was how the *Times* attempts to execute on them—specifically, where and how they reach their audience. This has required equal parts ruggedness and flexibility, and, as you'll see, it remains an ongoing challenge.

As early as 1994, the *Times*'s publisher, Arthur Ochs Sulzberger Jr., commented, "If they want [our content] on CD-ROM, I'll try to meet that need. The internet? That's fine with me.... Hell, if someone would be kind enough to invent the technology, I'll be pleased to beam it directly into your cortex." Smartphones aren't inter-cortical channels, but they're pretty close. As such, by 2010 the company had already made digital readership its top priority, well ahead of its competitors. It was also an industry-first mover when it created a paywall on its website in 2011. This was followed by a proliferation of subscription packages and products, including three tiers of news subscriptions as well as options for those interested only in crossword puzzles or cooking content. The varietal subscription model helped the *Times* become less dependent on advertising revenue. It also allowed the *Times* to pay for reporters, writers, editors, and producers. As advertising revenue continued to fall, not only in traditional print outlets but on the internet too, the *Times* focused on developing podcasts, with shows like *The Daily* and *The Ezra Klein Show*. The podcast network provided yet another channel for connecting with "readers," and it represented an offering in which advertising spend was still relatively strong.

The *Times*'s biggest challenge to date has been integrating all of its rapid differentiation, which requires more than a mere branding exercise. Here is the *Times* executive editor, Dean Baquet: "I always try to question the difference between what is truly tradition and core and what is merely habit. A lot of stuff we think is core is truly just habits. I think that's the most important part about leading a place going through dramatic change and even generational change. Here's what's not going to change—this is core, this is who we are—and everything else is sort of up for grabs."

Rather than try to insulate itself from change, as so many

other newspapers did, the *Times* saw itself in conversation with it. The *Times* has successfully adapted to date, but the future remains uncertain—because of course. Whether or not it will withstand the next series of external shocks remains to be seen. The biggest challenge for the *Times*, and for all news-media organizations, will be drawing a distinction between what constitutes news and what constitutes entertainment (as the late media theorist Neil Postman predicted back in 1985, we are increasingly "amusing ourselves to death"); figuring out the balance between deeply reported long-form stories, thoughtful essays, and more superficial but highly clicked-on content; and of course, deciding how to define the "truth" and uphold its core value of following it wherever it leads, even if it upsets readers.

Guiding Your Own Evolution

Central to the argument of Thomas Kuhn, the philosopher whose work we discussed briefly in chapter 1, is that scientific progress follows a predictable cycle of order, disorder, and reorder. Kuhn's masterwork, *The Structure of Scientific Revolutions,* contains a few sentences that I think carry outsize importance. Toward the end of the book, Kuhn explains how scientific crises eventually transition into new and stable paradigms. "In those situations where values must be applied, different values, taken alone, would lead to different choices . . . There is no neutral algorithm for theory-choice," he writes. To fully understand how science progresses through uncertainty, he goes on, one must understand the "particular set of shared values" held by the scientists trying to solve the problem. During periods of change and disorder—what Kuhn calls "crises"—a new and stable paradigm emerges not by chance, but as a result of the values held by the people who are doing the work. The scientists navigating uncertainty follow their values until they arrive somewhere new. Scientific progress is not random. It is directed by values, most notably, the scientific method, as we saw in the example of Robert Wilson, Arno Penzias, and the cosmic microwave background. The same is true for personal and organizational progress, too.

In the prior chapter, we learned to conceive of ourselves fluidly. We realized that developing complexity (differentiation and integration) is essential to thriving in relationship to our ever-changing environments. In this chapter, we learned that we don't grow complex and evolve at random. The ways in which we differentiate and integrate, the direction our paths take over time, owes itself to a combination of our rugged core values and our willingness and ability to apply them flexibly. Put all of this together and the result is a *rugged and flexible identity*.

Audre Lorde defied any single categorization. Lorde, who passed away in 1992, described herself as "black, lesbian, mother, warrior, and poet." She grew up in New York City and went to Catholic school before attending a public high school. She matriculated to Hunter College, where she majored in library studies. From there, she took a job as a librarian in the New York public schools. She and her husband, Edwin Rollins, a white man who eventually came out as gay, had two children together before they separated in 1970. In 1972, she met her partner Frances Clayton. All the while, she published poetry and prose that wrestled with topics such as gender, sexuality, race, and discrimination. As such, Lorde became an essential figure in numerous activist movements, ranging from civil rights, to feminism, to LGBTQ equality.

During a routine self-exam in her forties, Lorde found a

lump in her right breast. She got it biopsied and, thankfully, the result came back negative. But then, less than one year later, in September 1978, she went in for an additional screening. This time, the tumor was identified as malignant. She began reflecting on and recording her experience with cancer in journal entries and essays, which she eventually published as a book, *The Cancer Journals*. In it, she recalls her reaction to the news of the tumor's malignancy. "Off and on I kept thinking. I have cancer... Where are the models for what I'm supposed to be in this situation? But there were none. This is it, Audre. You're on your own," she writes.

The person she ended up becoming was an even more expansive version of herself, a person whose identity *included* death, and whose life became a part of the social movements she championed. "I carry death around in my body like a condemnation. But I do live," she writes. "There must be some way to integrate death into living, neither ignoring it nor giving into it." Her solution was using her mortality as fuel to work toward her core values with heightened vigor. Just as her life had been devoted to her values of justice and equality, so, too, was her death. "If I do what I need to do because I want to do it, it will matter less when death comes, because it will have been an ally that spurred me on," she wrote.

Lorde may have accepted and integrated death into her sense of self, but it does not mean she embraced it with open arms. She is candid throughout *The Cancer Journals* about the dread and despondency she felt as a result of her diagnosis. But Lorde took solace knowing that the fight for justice and equality did not begin with her birth and would not end with her death. By placing her life's work within a continuum of other activists, writers, and poets, Lorde adopted an interdependent perspective of her identity and, toward the end of her life, she leaned into an "ultimate" conception of herself. She considered herself a constituent of something larger and more enduring than her body, something that would carry pieces of her far into the future, long after she died. Given that the activism to which she devoted herself is still going strong and her writing is still widely read and cherished, it is safe to say she succeeded in this aim.

Lorde's writing brings to mind specific teachings of the recently deceased Zen master Thich Nhat Hanh. Hanh counseled that if we act in alignment with our core values, then we live on through the reverberations of our actions. He called this our *continuation body*. "We don't need to wait for the complete disintegration of this body in order to begin to see our continuation body, just as a cloud doesn't need to have been entirely transformed into rain in order to see her continuation body," he writes. "If we can see our continuation body while we're still alive, we'll know how to cultivate it to ensure a beautiful connection in the future. This is the true art of living."

Hanh taught that our actions are our only true belongings. We cannot escape the consequences of our actions. Our actions represent the ground upon which we stand. The importance of this cannot be overstated. In his 2022 book, *What We Owe the Future,* philosopher William MacAskill introduces the phenomenon of "early plasticity, later rigidity." During and immediately following periods of rapid change, there is a brief window to participate in the creation of a new normal. Over time, however, that window closes, and things calcify and become rigid again. This means that, particularly during periods of disorder, values-driven actions can have effects that last for centuries and beyond. Change and disorder may be uncomfortable, but they present an enormous opportunity to shape the future—be it of ourselves, our organizations, our communities, and even entire societies.

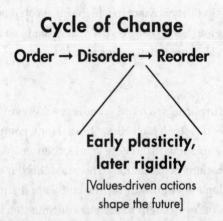

Even (and perhaps especially) if we don't know where the path ahead is going, we'd be wise to adopt an attitude of simply doing the next right (i.e., values-driven) thing. This gives us the best chance of getting where we ought to go. Developing rugged flexibility is anything but passive. Taking thoughtful and deliberate actions is what being in conversation with change is all about—and it is the topic we'll examine in the next, and final, part of this book.

Develop Rugged and Flexible Boundaries

- Your core values are the principles by which you live; they serve as the rugged boundaries of your identity, guiding how you differentiate, integrate, and navigate your path.

- It is good to have three to five core values. Define each in specific terms and come up with a few ways you can practice each in day-to-day life.
- When you feel the ground shifting underneath you, when you don't know your next move, you can ask yourself, *How might I move in the direction of my core values?* Or, if that isn't possible, you might consider, *How might I protect them?*
- Flexibility is about continually adjusting how you practice and apply your core values in ways that are true to yourself but also in harmony with your changing circumstances.
- It is normal for your core values to change over time. Navigating the world using your current core values is what guides you to discover your new ones.
- "Early plasticity, later rigidity" means that values-driven actions are particularly important during periods of change and disorder; they have outsize impact in shaping the future.

Part 3

RUGGED AND FLEXIBLE ACTIONS

5

Respond Not React

Over two thousand years ago, the Stoic philosopher Epictetus opened his handbook with the following line: "Of things some are in our power, and others are not." The remainder of his text elaborates on this dichotomy, laying out what is widely considered the most essential of all Stoic teachings: There are many phenomena in life that we cannot control—examples include aging, illness, an angry boss, the weather, poor reception to our work, our competition, and the missteps of our children. What we can control, however, is what we do in response, and that is where our focus ought to lie. Though the origin of this logic is widely attributed to Stoicism, which spread across the West between 200 BCE and 200 CE, a similar concept was already being pursued a few hundred years earlier and to the East. In the foundational Taoist text *Tao Te Ching*, published in 400 BCE, Lao Tzu wrote, "The master allows things to happen. She shapes them as they come."

It is not my objective (nor is it in my wheelhouse) to parse out whether this reasoning originated independently in the East and the West or spread from the East *to* the West. My interest here is simply that two principal ancient wisdom traditions coalesced around the same fundamental truth: we cannot control what happens to us, but we can control how we respond.

This truth has survived the test of the time. Perhaps the most well-known Christian prayer is that for serenity, published in 1951, by the theologian Reinhold Niebuhr:

> God, grant me the serenity to accept the things that
> I cannot change; courage to change the things I can;
> and wisdom to know the difference.

This truth has also survived the scrutiny of empirical science, featuring prominently in nearly all of the modern, evidence-based mental health therapies, including acceptance and commitment therapy, cognitive behavioral therapy, dialectical behavioral therapy, and mindfulness-based stress reduction. Each of these approaches teaches the importance of parsing out what you can control from what you cannot, and then learning how to focus on and take responsibility for the former while not getting caught up in or blaming yourself for the latter.

Our central topic here—change—is something we cannot control. It is an omnipresent, unpredictable force in our lives. The best we can do is learn how to dance with it, doing what we can to ensure our actions are as values driven and effective as possible. As you'll see in the coming pages, there is a science—and an art—to practicing this.

Katie[*] is a fourth-grade teacher at a medium-size public elementary school in western North Carolina. In March 2020, early in

---

[*] This person's name has been changed to protect her identity.

the pandemic, before vaccines and therapeutics were available, her school district transitioned to an exclusively remote learning model. She had three days to figure out how to convert her entire curriculum online. This was difficult for all teachers but especially those like Katie who work with younger students. It is hard enough to get a ten-year-old to pay attention for long periods of time in person. It is nearly impossible virtually, particularly when everyone is filled with questions and concerns about the state of the world and their own health and safety. When the stark transition to remote learning was first announced, Katie panicked for a few minutes. But then she quickly came back to what she could control about the situation, devising condensed lesson plans for the upcoming week. They were less than perfect, but they were something to get her and the kids started.

When it became clear that the initial COVID surge was not going to dissipate, Katie's district sent computers to all students. This was an important and welcome gesture, but it did not come without challenges. "A couple of the kids were missing in action. As I was trying to get a hold of them, I also had to figure out how to get everyone else onto Google Meet for the first time, which is, evidently, hard for fourth graders," remembers Katie. "Parents would call me for advice on how to get their wireless hot spot to work. Many of my colleagues were asking me how to use various online features. I became a technology coordinator at the same time I was trying to convert all my lessons to online formats for a bunch of young kids."

By the time the following school year began in the fall of 2020, many of Katie's colleagues had resigned, which led to teacher shortages and increased class sizes. This didn't add much extra hassle at first, since everyone was online. But in 2021, when the school transitioned to a hybrid model, the teacher shortage made an already messy situation even messier.

In a well-intended (but perhaps poorly thought through) effort to protect vulnerable children and family members, Katie's school district decided that in-person attendance would be optional. What this meant for teachers is that they had to teach to students in person and online simultaneously. Large companies spend thousands of dollars on supportive technology to host a two-hour hybrid meeting for high-functioning adults, including production and IT workers whose sole job it is to synchronize and troubleshoot. Katie, meanwhile, was given a half day, a new laptop, and a pat on the back—and then told to figure out how to host the equivalent of one hundred consecutive daylong meetings for a bunch of stressed-out fourth graders.

Hybrid teaching required endless improvisation, with no instruction manual to help. For instance, when Katie observed the emotional instability of her students, she responded by instituting regular mental health checks throughout the day, even if they came at the expense of time on traditional subjects such as math and science. She was constantly being forced to evaluate impossible trade-offs and make imperfect decisions, responding to an unprecedented and unpredictable situation as thoughtfully as possible, day by day, week by week. Prioritize the at-home students or the in-person ones? Prioritize math or mental health? Prioritize social support or literacy? The list goes on.

In 2022, when her school returned to an entirely in-person model, she faced new difficulties. "Many of my students hadn't had a full, normal school year since first grade, and it was a big and draining adjustment," she explains. On top of this, in a timing decision that seems utterly nonsensical (and that's putting it kindly), administrators decided to roll out an entirely new curriculum. This required Katie and her colleagues to recreate all of their lessons. Unsurprisingly, this precipitated more teachers quitting, which in turn increased Katie's class size yet again, making the social distancing they were supposed to be practicing impossible. Her first thought was to open the windows in order to increase airflow, but the vast majority of them were jammed and broken. Instead of panicking or becoming engulfed by rage, she responded creatively. "If you walked into my classroom you'd see one of my windows propped open with a giant fossil rock because there's no way to get it to stay open otherwise," she told me. "I had to do the best with what I had."

When I asked Katie why and how she kept showing up throughout the chaos, her response was twofold. First, she tried to reflect on her core values every day. "I come back to my why and what I do this for. It's not for the district or the superintendent. It's for the kids," she told me. Second, she acted in alignment with those values by responding to the things she could control, and she tried to let go of everything that she could not. "Some teachers take every mandate from the top super seriously and then they just get really frustrated and give up or quit," she explains. "I feel a loyalty to the kids; I care about them, and I keep them in mind when we're in a meeting and we're told that we have to do something that seems completely ridiculous. I'll just nod and then when I get in the classroom I do what I think is best," she explains. "You can spend forever and a day getting stuff ready and then something happens and your whole day is thrown off anyway. You have to do what you can do in the moment to create a positive environment for the kids, just constantly responding to whatever is in front of you, which during COVID, is changing all the time."

In addition to demonstrating that teachers are spectacularly underappreciated and undersupported in just about every way, Katie's experience illustrates the power of focusing on what you can control, and not worrying about what you cannot. She encountered disorder and chaos not by reacting rashly, panicking, or going on autopilot mode, but by responding with intentionality instead, by thoughtfully doing the next right thing.

Zanshin

An extreme (and literal) example of going on autopilot is something called *target fixation*, a phenomenon most frequently observed in drivers, motorcyclists, and pilots. Defined broadly, target fixation is when a person becomes so focused on a particular target they are headed toward that they end up driving, riding, or flying right into it. The most common example occurs when a driver focuses too closely on the car directly in front of them, only to end up ramming into its rear end. Another example is crashes on the "shoulder." These commonly occur when drivers identify a car that is pulled off to the side of the road, channel their attention toward it, and then proceed to wallop into it.

While the research on target fixation is limited to mechanical transportation, I suspect its theme holds true in all walks of life. If we become too focused on the next thing in front of us, then we risk thoughtlessly crashing into it. It is the mountain climber who is so fixated on reaching the peak that they overlook subtle cues foreshadowing changing weather, and thus expose themselves to unacceptable risks on their descent, an occurrence colloquially referred to as *summit fever*. It is the parent who is so worried about their child's future achievements that they overlook what their child needs right now. Or the manager who so desperately wants to get promoted that they end up underperforming on the work in front of them. Zoom way out and target fixation becomes a hazard for existence as a whole. If we are perpetually latched onto something out in front of us, to our future plans and where we think we are headed, then we run the risk of pummeling directly into the ultimate target—our deaths—without knowing how we got there. We miss out on all sorts of interesting stuff along the way.

The martial art aikido recognizes the problem of target fixation and counters it by teaching a quality called *zanshin*, defined loosely as continuing awareness that prepares you for your next action. Zanshin focuses on what is happening in front of you, but also on what is happening around you. It is a flexible way of seeing that zooms in and out and rotates this way and that. In contrast to target fixation, zanshin allows for the perception of an object or goal and the field surrounding it at the same time. It may begin as an asset for aikido, but following its own prescription, zanshin extends beyond any single aim. "Zanshin is the future, but zanshin is also now. The quality of your zanshin is the quality of your aikido, and the quality of your aikido is the quality of your life," writes the aikido master and humanistic philosopher George Leonard.

I Drive Safely, one of the largest driving schools in the United States, approaches the difficulty of target fixation by teaching zanshin, even if they don't call it that. "If an unexpected object enters your vision, such as a car pulling out, a squirrel running across the road, don't look at it directly—use your peripheral vision and plan to look beyond the object in question." This softening of your view allows you to relax into whatever is happening and pick up on delicate cues that you might otherwise miss, which in turn allows you to respond to an unfolding situation more effectively.

Whether you are driving a car, competing in aikido, parenting a child, or managing a team, zanshin's power is that it places you in conversation with your changing circumstances. When you catch yourself fixated on a single end point—thinking about it nonstop, perhaps even tensing in your body—ask yourself what it would look like to zoom out and consider not just your goal or objective but what is happening around it too. Pause and contemplate alternative paths to getting where you want to go, and consider that what you think is your final destination may not actually be your final destination at all. (A four-step method can help, and we'll unpack it later in this chapter.) Regardless of what life throws your way, by practicing zanshin you give yourself a better chance at deliberately responding to changing circumstances instead of going on autopilot and reacting. Your actions become more refined and aligned with your present-moment reality and your values instead of with false perceptions or prior expectations. As a result, you not only feel better, but you do better, too.

The year is 2008. The setting is the Interlachen Country Club in Minnesota. The US Women's Open golf championship is unfolding in a manner no one could have predicted. A nineteen-year-old from South Korea is dominating, sinking challenging putts from all over the greens. Her name is Inbee Park, and it is only her second year on the professional tour. She is so far ahead that by midway through the tournament's final round, it is clear that she will make history, becoming the youngest player to capture the prestigious title.

Since winning the US Open in 2008, Park has gone on to have an astounding career. She's been ranked as the top woman player in the world on four separate occasions. She's won twenty-one tournaments, including seven major championships and a career grand slam, becoming only the fourth woman in history to win all four of golf's crown jewels. And in the summer of 2016, despite a significant hand injury, Park won gold at the Rio Olympic Games.

Park's tremendous success owes itself largely to her zanshin. She is known for staying calm and collected regardless of what is happening, allowing her to skillfully navigate change and challenge. This is perhaps most apparent in her putting, where Park's excellence is unrivaled, irrespective of gender. The golf writer Max Schreiber explains that she is "making putts from 10–15 feet a remarkable 64% of the time... For context, the rest of the LPGA [the women's division] averages 28% from that range, while on the PGA Tour [the men's division] it's about 30%... The longer the putt, the more relaxed and confident Park is."

Park attributes her proficiency on the greens to her relaxed attention and her ability to focus on what she can control while letting go of the rest. "There are so many variables that you have to take into account... So many things happening on the way to the hole.... I'm just trying to putt the right speed, the right line, and that's pretty much all I can do," she told the GOLF channel.

In golf, putting is colloquially referred to as the "short game," since you are relatively close to the hole when you strike the ball. But Park's zanshin extends to how she approaches the long game, too—and not only on the fairways. Following her breakout championship at age nineteen, Park experienced a rough patch where she struggled with the pressure of high expectations. She got ahead of herself and became worried about winning whatever tournament was next on her schedule, losing her ability to respond to what was happening in the moment. In the end, she relied on zanshin to carry her through. "I came to realize that [expectations] really don't matter and people don't care about you that much, as much as you think... You don't really need to worry about other people. Worry about yourself and that what you're doing is right," she says, and the subsequent moves will emerge on their own. We'll now peer under the hood to see why, examining fascinating neuroscience that supports responding instead of reacting and doing the next right thing.

Behavioral Activation and the Neuroscience of Responding Instead of Reacting

When we are confronted with change, especially if it is sudden, a part of our brain called the *amygdala* livens with activity. The amygdala is an older structure, evolving early in the history of *Homo sapiens*. Its primary purpose is to make us kick, scream, and run if we are being attacked by predators. It is a central part of what the late neuroscientist Jaak Panksepp called the *RAGE pathway*, or the neural circuitry that is predictably activated when our sense of self and stability come under threat. The RAGE pathway evolved to be reactionary, and for good reason. If we are being chased by a lion or tiger on the savanna, our survival depends upon quick and instinctive moves. In today's world, however, our threats rarely come from lions or tigers, and thus they don't require as much of an immediate reaction. If anything, reacting to modern disruptions—such as climate change, aging, illness, workplace tensions, and relational disputes—tends to backfire. Most of our current challenges require thoughtful and deliberate responses instead.

To be sure, the RAGE pathway can still be useful, especially if you spend time hiking in bear country. It's just that its utility is limited to far fewer circumstances than it once was early in our species' evolution. Fortunately, we are equipped with other options.

Another brain region, the *basal ganglia*, receives direct input from the amygdala via a group of neurons that make up a tiny structure called the *striatum*. You can think of the striatum as a multi-lane highway connecting the basal ganglia to the amygdala, as well as to other parts of the brain. The basal ganglia is not solely concerned with RAGE. It controls other behaviors too, including what Panksepp named the *SEEKING pathway*. The SEEKING pathway facilitates planning and problem-solving. It underlies our ability to exert agency and consciously move toward challenges instead of falling into helplessness or impulsively running away. Much of Panksepp's groundbreaking research took place in a field called *affective neuroscience*, which, among other things, aims to connect specific brain networks to the emotions and behaviors they manifest. His work (and that of other investigators) shows that the RAGE and SEEKING pathways compete for resources, engaging in what amounts to a zero-sum game: if the SEEKING pathway is turned on, then the RAGE pathway is turned off.

What Panksepp and his colleagues made visible using cutting-edge neuroscience is something to which most of us can relate in our own lived experience. When we are making plans, problem solving, or deliberately working toward a challenge, it is nearly impossible to be raging and filled with anger at the same time. The brain is incapable of responding and reacting in parallel, and by engaging the features that make up the former, we prevent ourselves from spiraling in the latter. But the story does not end there.

The neural circuitry associated with responding is like a muscle: it gets stronger with use. Neurons that fire together wire together. If you are able to muster a deliberate response in a difficult and distressing situation today, you'll be more likely to do so out of habit tomorrow. With each calculated action we take, a spurt of the neurochemical dopamine is released. Dopamine acts as fuel for the SEEKING pathway. It makes us feel good and motivates us to keep going, even if we are walking along challenging and uncertain paths. The more fuel the SEEKING pathway has on which to operate, the less likely the RAGE pathway will overtake it.

In his fascinating book on the origins of consciousness, *The Hidden Spring*, South African neuroscientist Mark Solms writes about what he calls the *Law of Affect*. The Law of Affect says that while our thoughts are undoubtedly important, it is predominantly our feelings—our affect—that dominates our consciousness and thus directs us this way or that. As such, we tend to repeat behaviors that make us feel good. The SEEKING pathway, and the dopamine that fuels it, is implicated in many of these behaviors—be it planning or taking micro steps to exert agency and achieve goals. The result is a virtuous cycle: if we deliberately respond to uncertain situations, we feel good, and we become more likely to deliberately respond again. Remember, this is so important because the RAGE pathway is automatically turned off when the SEEKING pathway is turned on. Once we get into a productive groove, our brains are less likely to be hijacked by hot and potentially destructive emotions.

Because we are fallible humans, try as we might, anger, panic, and other reactive emotions will, on occasion, get the best of us. The good news is that the brain has a built-in defense mechanism for when this happens. The bad news is the defense mechanism is that we become quite sad.

The RAGE pathway can only be active for so long before it is exhausted. When this happens, another set of neural circuits, some neuroscientists call it the *SADNESS pathway,* comes online. The result is despondency and dejection. Many of us have experienced this. For instance, when we lose our cool with a partner, friend, child, or colleague it might feel good in the moment—*finally, it was SO nice to let loose and explode*—but following such explosions most people feel bad. (It brings to mind the Buddha's poignant description of anger as a "poisoned root with a honeyed tip.") Meanwhile, if a longer period of anger, rage, or the other common reactive emotion, panic, is left unresolved, there is a high likelihood of someone sliding into burnout, chronic fatigue, and even clinical depression.

A part of what makes mind states like helplessness, despair, and depression so insidious is that they quickly become entrenched and thus make it extremely hard to galvanize the SEEKING pathway. "Chemically, the transition from 'protest' to 'despair' is mediated by peptides which shut down dopamine. This is why depression is characterized by the mirror opposites of the feelings that characterize seeking," writes Solms. I suspect this is also why behavioral therapies are often more powerful than cognitive ones. Whereas with cognitive therapies you try to *think* your way into a new state of mind, with behavioral therapies you try to *act* your way into a new state of mind, even if it feels like you are forcing yourself. But if you can muster the will to take one small and productive action today, you engage your SEEKING pathway and make it more likely that you will be able to take one small and productive action tomorrow. Dopamine is released, and productive SEEKING behavior builds on itself.

You may be wondering, *Isn't planning a form of thinking? And didn't you previously write that planning can activate the SEEKING pathway as well?* The answer to both questions is yes. But even the most optimistic or strategic thinking is not associated with the same bolus of dopamine that acting is. This is why taking productive action is particularly helpful when you are feeling down, unmotivated, or apathetic—when your brain's SADNESS pathway is dominant. You can give yourself permission to feel those feelings but not dwell on them or take them as destiny. Instead, you shift the focus to taking just one action, bringing your feelings, whatever they may be, along for the ride. Doing so gives you the best chance at improving your mood. Clinical psychologists call this *behavioral activation,* and it is based on the idea that action can *create* motivation and positive affect, especially when you are in a rut. In layperson's terms, you don't need to feel good to get going; you need to get going to give yourself a chance at feeling good.

It can be helpful to think of the initial oomph to get going as *activation energy*. Sometimes you need more, and sometimes you need less. Productive actions are self-reinforcing. The more you can nudge yourself to get going today, the easier it becomes tomorrow. To be sure, behavioral activation is not a be-all and end-all for people experiencing depression, despair, or other mental health challenges. But clinical research shows that it can be an extremely effective tool, especially when you are feeling down and dispirited during periods of change or disorder.

If you don't know where to begin, a good place to start is by reflecting on your core values, which we discussed in the prior chapter. Then ask yourself how to apply your activation energy strategically. What actions work in service of your core values and will give you the jolt you need? You may not feel like getting started, but get started anyway, and see what happens. Though we are conditioned to think that our being influences our doing, it is amazing how much the opposite is also true: our doing influences our being, too.

All of this fascinating neuroscience essentially warns us about freaking out when there is a change and then, after a period of panic and/or anger, spiraling into hopelessness, fatigue, and despair (which is not uncommon). It shows us why deliberately responding with values-driven actions is a much better, albeit harder, path. We'll continue to explore this path in the following pages.

The Extraordinary Power of Responding Not Reacting

Cristina Martinez was born in Capulhuac, Mexico, where her family made the quintessential fare of the region: barbacoa, which involves slow-roasting meats over an open fire in a marinade of lime juice, olive oil, and other local seasonings. Growing up, Martinez worked for the family food service, coming to know the cuisine intimately. She considers this to be a relatively stable and happy time in her life.

At seventeen, she married a man whose family also made barbacoa. However, they forced her to work brutal hours, laboring from three in the morning until past ten in the evening. When she finally told her husband the work was unsustainable, he began to treat her terribly, verbally and physically abusing her. She took solace in raising her daughter, Karla, whom she considers the light of her life. Martinez didn't want Karla to experience the same abuse and lack of autonomy that she did. She desperately wanted Karla to receive an education that would enable her to pursue her own career.

In order to get a reputable education, Karla would have to go to boarding school, which required Martinez to make extra money for tuition. Unfortunately, her husband kept every penny. "When Karla was barely thirteen, her dad told me, *I hope to make a good wife out of my daughter.* I took those words to heart and thought, 'Wow. I am about to lose my daughter,'" remembers Martinez on the Netflix special *Chef's Table*. "I said to Karla, 'I don't want you to repeat this same story.'"

It became clear that Martinez would need a way to earn income that was separate from her husband and his family's business. Left with no other choice, she responded by deciding to leave her home and family and immigrate to the United States. Her plan was to find work in Philadelphia, where she had a brother-in-law, and then send whatever money she earned back to Karla, whom she would move to safety. Martinez found a local "coyote" to help her cross the border. They'd leave in a few months. Martinez trained by running every day and made sure to eat well so that she had enough energy for the long and arduous journey.

On the appointed day in 2006, Martinez and twenty-three other people flew to Juarez City and began walking. They walked for fifteen days in the desert, battling the elements and surviving on meager rations, but they made it. After a seven-day drive across America in a car that her smuggler had arranged for her, Martinez arrived in Philadelphia. She immediately began looking for work in kitchens and was hired to do food prep at an Italian restaurant.

She didn't know English and the chef didn't know Spanish, so she had to learn by observing, which she did as if her life depended on it, which in her case, it did. She excelled and was quickly promoted to pastry chef. In the kitchen, she met the man who would become her second husband, Benjamin Miller, an American coworker. After dating for many months, Miller proposed to Martinez and they got married soon after.

Following their wedding, Miller tried to help Martinez attain her green card but the lawyers said they needed a letter from her employer. When they asked the restaurant's owner, he denied them, claiming not to have known she was undocumented. He promptly proceeded to fire her. Without work, Martinez suddenly had no way of sending money home to Karla. Instead of reacting with rage, she thought about what she could do and noticed that while many immigrants from Mexico lived in her neighborhood, there was no barbacoa establishment. And therein lay her calculated response. "There were many options on the local menus—seafood, meat, and traditional dishes—but no barbacoa. So I thought: 'Maybe I can sell barbacoa here,'" she recalls in an interview about her life.

Martinez started cooking barbacoa in the apartment she and Miller shared. Miller helped her make business cards and handed them out to anyone who would take one. When they realized they could not get authentic ingredients nearby, they responded by establishing a relationship with a farmer in Lancaster, Pennsylvania, so that they could grow their own. People fell in love with her food; those who had immigrated from Mexico became emotional when they tasted it because it reminded them of home.

Her cooking became so popular that she needed to find a space outside of her and Miller's apartment for a proper kitchen. Luckily, a friend who owned a restaurant was moving out of the space, and Martinez could take her operation there. Miller helped run the business, and Martinez cooked with a ferocious intensity and joy like never before. Their restaurant continued to increase in popularity, becoming the epicenter of a fast-growing immigrant community in South Philadelphia. In 2016, the popular culinary magazine *Bon Appétit* named her neighborhood joint one of the top ten new restaurants in America. "All of a sudden, I'm on the radio, on TV, in magazines," Martinez remembers. She responded to the publicity by sharing her story and advocating for the rights of immigrant workers. In addition to amplifying her voice, the publicity also increased the popularity of her food. Today, Martinez acts as an outspoken activist for immigrant labor reform, and her restaurant, South Philly Barbacoa, serves more than 1,500 patrons every weekend. All the while, she funded a proper education for her daughter, Karla, who is currently a nurse and able to support herself.

Immigration is a complex issue; but at the level of an individual person, I hope we can agree it should not be so hard for any well-meaning person to secure their safety and dignity. As with Katie's story from earlier in this chapter, and Bryan Stevenson's from chapter 2, it is unfortunate that people are forced to turn to such heroic measures to begin with. But here we are. Recall what I wrote in the previous chapter: if we are to have any chance at improving a broken world, we must learn how to navigate it without becoming broken people. Here, there is much to learn from someone like Martinez, who was able to work through extreme disorder without falling into despair. She did this by responding instead of reacting to the dramatic challenges and setbacks she faced.

There were several moments throughout Martinez's journey during which she could have easily been overtaken by her RAGE pathway, but she engaged her SEEKING pathway instead. When she was trapped in an abusive relationship with her first husband and saw the love of her life, her daughter Karla, heading toward a similar future, Martinez began making plans to immigrate to the United States so she could send money back. When she asked for her boss's help in obtaining her green card and he reacted by firing her, she could have easily fallen into RAGE—especially because she had worked so hard for his restaurant. But her response was to get curious about gaps in her local neighborhood's culinary scene, which led her to making barbacoa. When she couldn't get enough of the right ingredients, she sought out local connections to grow them. We can observe the same pattern in the fourth-grade teacher Katie's story, who kept showing up and responding as thoughtfully as possible during a period of difficulty, change, and disorder.

No doubt, they might have been full of anger at times, but neither of these women gave their RAGE pathways the opportunity to fully seize their consciousness. If anything, they turned their anger into fuel for productive action. They separated what they could control from what they could not, and then they focused on the former, asserting their agency by responding instead of reacting. Do this repeatedly and you begin to develop what psychologists refer to as self-efficacy, a secure confidence borne out of the evidence-based belief that you are capable of showing up and taking deliberate actions during difficulties. Decades of research show that individuals who score high on measures of self-efficacy are better able to work through periods of change and disorder. It makes sense. If you are insecure about your ability to respond to change, then you are liable to perceive a need to control everything and change becomes threatening. When things feel out of control, you default to reacting. But if you are secure in your ability to respond to change, then you become increasingly at peace with it, and thus more likely to skillfully navigate whatever life throws your way.

The 4Ps: An Evidence-Based Method for Self-Efficacy During Change and Disorder

Just because you know something intellectually doesn't mean you'll consistently act on it, particularly in charged situations. A common topic of discussion with my coaching clients is not only the benefits of responding instead of reacting, but *how* to actually do it. As such, I've developed a heuristic to help: 2Ps versus 4Ps: when we react, we *panic* and *pummel* ahead; when we respond, we *pause, process, plan,* and only then *proceed.*

Reacting is quick. You feel and then do. Responding is slower. It involves more space between an event and what you do, or don't do, about it. In that space, a *pause,* you give immediate emotions room to breathe and thus you come to better understand what is happening—that is, you *process*. As a result, you can reflect and strategize using the most evolved and uniquely human parts of your brain to make a *plan that is in alignment with your values,* and then *proceed* accordingly. Responding is harder than reacting, especially at first. It requires more psychic energy; it demands letting an urge to immediately do something, anything, be there without giving in. But like most things that require effort, responding tends to be advantageous, for all of the reasons we've already explored. You rarely regret deliberately responding to a challenging situation, whereas you often regret automatically reacting to one.

Pause

Just about anyone can pause for an instant. But when emotions are running high, it is all too easy to become overwhelmed, spiraling back into reactivity after only a split second. Genuinely processing a situation, particularly a challenging one, requires time and space. A powerful way to create it is by naming what you are feeling.

In a series of studies out of UCLA, researchers subjected participants to unplanned and distressing situations, such as giving impromptu speeches in front of strangers. Half of the participants were instructed to feel and label their emotions: for instance, "I feel tightness in my chest," "I feel angst in my throat," or "I feel heat in my palms." The other half were not instructed to do anything special. The participants who felt and labeled their emotions, what researchers call *affect labeling,* had significantly less physiological arousal and less activity in the amygdala, the part of the brain associated with reacting (and the RAGE pathway). The affect labelers also reported subjectively feeling more at ease during their speeches. Quite important to point out is that people who deeply felt their feelings but did not label them actually had *more* angst. In other words, it is the act of labeling that creates space between stimulus and response. The common aphorism to "feel your feelings" may only work if you also name them.

I suspect this is because if you simply experience what is going on, you are likely to get overly involved in that experience, perhaps even fusing with it. Deeply feeling anxiety, despair, or nervousness is no fun. But by labeling these emotions, you separate yourself from them; you come to *know* what you are experiencing instead of simply experiencing it. Sometimes referred to as meta-awareness, this backward step in perception affords you more freedom to process whatever is happening.

Research on affect labeling is less than a decade old. But the idea dates back hundreds of years to a concept prevalent in ancient mythology and folklore. Called the *law of names,* it states that knowing something's true name—you can't just be close; you've got to hit the nail on the head—gives you power over it. For example, in Scandinavian folklore, magical beasts could be defeated by calling out their true names. The Norwegian legend of Saint Olaf recounts how a saint was coerced and captured by a troll. The only way the saint could free himself was if he learned the troll's true name. In arguably the most widely known example, the German fairy tale of Rumpelstiltskin, the female protagonist owes her firstborn child to a villain who will only give up his claim under one condition: if she can guess his true name. (Spoiler alert: it's Rumpelstiltskin.)

Perhaps we should not be surprised, then, that in yet another case of convergence between modern science and ancient wisdom, the UCLA researchers also found the more granular someone's naming of an emotion—for instance: "longing" instead of "sadness," or "tightness" instead of "anxiety"—the better they are able to respond to both the emotion and the situation that gave rise to it. Knowing something's name really does give you power over it, and the more precise, the "truer," your naming is, the more power you have. With that additional power comes additional space, and with additional space comes additional autonomy and self-efficacy to respond instead of react.

Process and Plan

Once you've named an emotion and created space between you and the situation that gave rise to it, the next steps are to process and make a plan. Here, a handful of concrete psychological strategies can help. The first is practicing what the meditation teacher Michele McDonald calls *RAIN:* *R*ecognize what is happening. *A*llow life to be just as it is. *I*nvestigate your inner experience with kindness and curiosity. *N*on-identify with your experience, viewing it from a larger perspective instead. When you zoom out and view situations from a larger perspective, your ability to work with them in a skillful and responsive manner improves. Research shows that this is true for everything from physical pain to emotional pain to social tension to making difficult decisions. What follows are a few ways to cultivate this sort of larger perspective.

When you find yourself confronted with uncertainty and change, imagine that a friend or colleague is in the same situation as you. Visualize deeply that they are going through what you are. How would you look at that friend? What advice would you give them? Studies out of the University of California, Berkeley, show that this method helps people see clearly and respond wisely during all manner of circumstances, particularly when the stakes are high. You can also imagine an older and wiser version of yourself, perhaps ten, twenty, or even thirty years down the road. Maybe future-you is sitting in a cozy library and sipping on bourbon or tea. Or maybe you've got your grandchildren or lifelong friends over. What advice would older and wiser future-you give to your current self? What would it look like to follow that counsel right now?

The above strategies fall under the umbrella of what psychologists call *self-distancing*. Its purpose is to create the space and collected state of mind so that you can see clearly what is happening (process) and come up with subsequent actions (plan). In addition to being beneficial in moments when the SEEKING and RAGE pathways are vying for control, these exercises are also helpful over the long haul. Each time you self-distance, you cultivate zanshin and a perspective that is broader, more robust, and more enduring than your moment-to-moment, always-changing experience. As such, you strengthen secure self-efficacy and self-confidence, which, as we learned earlier, builds comfort with change. All of these concepts—zanshin, self-efficacy, responding, SEEKING, affect labeling, and self-distancing—work together to bolster rugged and flexible actions.

Meditation can also help you develop the ability to respond not react. Unlike common portrayals in the West, meditation is not about achieving a relaxed, blissed-out state. Rather, it is about learning to sit with various thoughts, feelings, and sensations without reacting to them. Every time something that inclines you to react arises—every itch, be it a physical or psychological one, that you so desperately want to scratch—the practice is simply learning to sit with it. Over time, you develop an attitude of curiosity toward whatever life throws your way, both on the meditation cushion and off it. You can be with, watch, and take interest in the thing instead of immediately reacting to it. You also develop compassion for yourself and for others, realizing how hard it can be to maintain a measure of equanimity amidst the torrent of thoughts, feelings, and urges that occur during just fifteen minutes of formal meditation, let alone a decades-long life. Meditation also helps you to not always take your conventional self so seriously. It supports your moving up Loevinger's ladder of ego development, strengthening the self that can laugh at, and perhaps even let go of, itself.

Still another way to process and plan in the midst of uncertainty is to experience awe—be it by spending time in nature, listening to moving music, observing poignant art, or a multitude of other ways. If reacting and the associated RAGE pathway represent, as Aldous Huxley wrote, a "reducing valve" of awareness, then awe helps us to open back up. Dacher Keltner, a professor of psychology at the University of California, Berkeley, has shown that awe is tied directly to feelings of spaciousness. Awe doesn't only improve the way we perceive and think; it enhances our biology too. According to a 2015 study in the journal *Emotion,* awe, more than any other sensation, is linked to lower levels of a molecule called interleukin-6, which is associated with stress and inflammation.

Unfortunately, we are increasingly awe deprived. "Adults spend more time working and commuting and less time outdoors and with other people," writes Keltner in a 2016 essay. He goes on to write that we've become "more individualistic, more narcissistic, more materialistic, and less connected to others." Is it really any surprise, then, that so many of us struggle to respond instead of react? If we never give ourselves the chance to create space in structured circumstances, how can we expect ourselves to create space in unstructured ones? If Keltner's argument is correct, which I believe it is, then a weekly walk in nature does wonders for our ability to respond instead of react in our day-to-day lives. I've experienced this firsthand. When I am struggling to come to grips with a big change or some underlying uncertainty, there is nothing like a long walk out of doors to help me more skillfully process what is happening and make subsequent plans.

Proceed

When researchers put mice in a maze and measure what happens in their brains when they accomplish micro-objectives on the path to distant goals (e.g., making a correct turn), what they find is that the mice's brains release dopamine, the neurochemical associated with motivation, drive, and the SEEKING pathway. But when they are given a compound that completely blocks their production of dopamine, the mice become apathetic and give up. Although these studies cannot be safely replicated in humans, scientists speculate we operate in the same way. The neurochemistry of progress primes us to persist.

It is much easier to get going in reacting mode than it is in responding mode. That's because reacting is instinctive. You just do. The problem, as I pointed out, is that "just doing" doesn't always give rise to the best next steps. When you respond, you give yourself the time and space to move forward more thoughtfully. The issue is that you also give yourself time and space to question whatever approach you've come up with, and you can all too easily fall into paralysis by analysis or become crippled by doubt.

The best antidote to these varieties of friction is to consider your actions as experiments. In the moment, there is no such thing as a right or wrong decision, so long as you've made it deliberately. If hindsight proves your actions useful, you keep going down the same path. If hindsight proves them unsuitable, you adjust course, perhaps repeating the first 3Ps—pause, process, and plan—before proceeding again. Every time you go through this cycle you are strengthening your SEEKING pathway, diminishing your RAGE pathway, and thus setting yourself up to become the kind of person (or organization) that meets change and disorder by responding instead of reacting. Research in continuous improvement from across diverse fields shows that going through loops like the 4Ps leads to the best outcomes during periods of change and disorder.

The Medium Is the Message

In 1964, the Canadian communication theorist Marshall McLuhan opened his book *Understanding Media: The Extensions of Man* by writing, "The medium is the message." He went on to explain that "the personal and social consequences of any medium—that is, of any extension of ourselves—result from the new scale that is introduced into our affairs by each extension of ourselves, or by any new technology." In layperson's terms: the more we use or consume a given technological medium, the more we come to represent it in our actions; or, unfortunately in today's day and age, our reactions.

Two of the most popular places we go to get information in the midst of change and disorder are social media and cable news. While both of these mediums can be effective sources for breaking news—that is, quickly telling us that such and such happened— their value beyond that is questionable at best. Rather than slow, deliberate, and thoughtful analysis, social media and cable news are dominated by decontextualized hot takes and people yelling back and forth at one another, or worse, yelling into the abyss. (Maybe this is actually better; I don't know?) Nearly everything about social media and cable news teaches us to react instead of respond. On cable news, nuanced and important topics are given a few minutes, at most, before the program moves on to the next thing. During those short segments, it is common to bring on pundits who are prescreened precisely for their likelihood of creating drama and rage. On social media, meanwhile, commentators are given a limited number of characters to make a point. Moreover, research shows the two factors that contribute most to the odds of a post going viral are the speed at which it goes up and the amount of outrage it stokes. Reactivity is incentivized and rewarded.

Brain science shows neurons that fire together wire together. The more you engage in certain patterns of thinking, feeling, and acting, the stronger those patterns become. It is hard to think of two better places to wire the RAGE pathway than social media and cable news. If the medium is the message, then these two mediums send a resounding message of reactionary impulses. Fortunately, there are plenty of other mediums that build up one's capacity to respond instead. A few examples include reading a book, having a distraction-free discussion with people whom you respect, or, if you are going to be on the internet, reading longer-form writing and not suffering any fools when curating your social media feeds.

It's worth reiterating that there is nothing inherently wrong with becoming aware of news on social media or cable television, so long as you check the integrity of your sources. But hanging around beyond that for so-called "analysis" quickly becomes deleterious. The challenge is that these mediums are intentionally designed to suck us in. After all, capturing our attention, which they monetize, is their core aim. The 2Ps versus 4Ps heuristic helps. You can always ask yourself, *Does this medium incentivize panicking and pummeling ahead or does it encourage me to pause, process, plan, and proceed?*

If you are swimming in a sea of reactivity, then it is inevitable that you will become a reactionary kind of person. If, however, you surround yourself with responsiveness, then you are liable to become that kind of person too. This is true not only individually but societally as well.

In this chapter, we discussed how rugged flexibility requires thoughtful engagement with the ever-changing tides of life. We learned about the difference between the things you cannot control and the things you can. We explored the benefits of focusing on the latter, and how a broader awareness called zanshin can help you to skillfully and deliberately respond to change and disorder instead of rashly and automatically reacting. We examined the neuroscience of both responding and reacting, saw how the SEEKING and RAGE pathways compete for resources, and how by activating the former we diminish the latter. We also learned that when the RAGE pathway is exhausted, the result can be despondency and depression; in these cases, more powerful than trying to think your way to a new state of mind is acting your way to one. We detailed an evidence-based method of responding to change and disorder and strengthening self-efficacy—the 4Ps: pause, process, plan, and proceed—and we learned about concrete tools for each element of the progression. Finally, we discussed how the mediums from which we consume information about change and disorder shape whether or not we respond or react in our own lives. If we want to be the kind of people who skillfully respond to change and difficulties, we should spend more time engaging in slower, responsible, and discerning mediums and less time with hot, fast, and reactionary ones.

We are going to end this chapter with an uncomfortable truth, but in the next one I'll provide some solace. Whether we like it or not, sometimes change and disorder shake us to our cores, and then some. There are certain circumstances during which we can do everything discussed in this book and yet still feel down, depressed, burned out, and void of growth and meaning. Perhaps we are up against something so big and overwhelming that there is little we can control (or at least it seems that way at first) and therefore even skillfully responding feels futile. At one point or another, we all go through these dark patches. They are inevitable parts of the human experience.

The next chapter is about what to do when these inevitabilities arrive in our own lives, when we find ourselves in the poet Dante Alighieri's infamous dark wood, "where the direct way is lost," where it is hard "to speak of how wild, harsh, and impenetrable that wood is." In these circumstances, trying to make sense of what is happening is often counterproductive. Sometimes the work of rugged flexibility is simply showing up and getting through. Meaning and growth cannot be forced. They must come on their own time. Fortunately, as we'll see in the coming pages, if we can learn to get out of our own way, they almost always do.

Respond Not React

- During periods of change and disorder, separate what you cannot control from what you can, and then focus on the latter while trying not to waste time and energy on the former.
- Becoming fixated on any given path or outcome often yields suboptimal results; instead, work on developing zanshin, or a broader, more curious, and more inclusive awareness.
- The best way to shift from the RAGE pathway and reacting to the SEEKING pathway and responding is by practicing the 4Ps:
    → *Pause* by labeling your emotions
    → *Process* by practicing non-identification, viewing your situation with remove
    → *Plan* by self-distancing and gaining even greater perspective as you evaluate your options
    → *Proceed* by taking micro-steps, treating each as an experiment and adjusting as you go
- If you get into the habit of responding instead of reacting, you develop what psychologists refer to as *self-efficacy*, a secure confidence borne out of the evidence-based belief that you are capable of showing up and taking deliberate actions during change and difficulties. The more self-efficacy you develop, the less threatening change and disorder become.
- The mediums from which you consume information shape your temperament; prioritize responsive ones and avoid reactive ones—your health, and perhaps that of society, depends on it.

6

Making Meaning and Moving Forward

In 2017, I was blindsided by the stark onset of obsessive-compulsive disorder (OCD) and secondary depression. OCD is a misunderstood and often debilitating disease. Far from a tendency to be meticulously organized or double-check that the door is locked or the toaster is unplugged, clinical OCD is characterized by intrusive thoughts and feelings that dominate your life, cratering your mood and warping your sense of self. You spend every waking hour trying to decipher what the intrusive thoughts and feelings mean and how to make them ease, only to have them come back stronger and more violently. They cause electrifying shots of anxiety from head to toe. You compulsively try to distract yourself from the intrusive thoughts and feelings, but they are always lurking in the background, exploiting any open space in your day, like a tab on a computer screen that you can never minimize, let alone close. You go to bed with intrusive thoughts and feelings crawling through your mind and body and you wake up the same way. They are there when you are eating. They are there when you are working. They are there when you are trying to be present for your family. They are even there when you are sleeping, tormenting your dreams. The intrusive thoughts and feelings are so persistent that you start to question whether you might believe them. It was a chaotic and bottomless spiral of pain and terror: this was my day-to-day reality for the better part of a year, before I began to notice the positive effects of therapy and other practices that have changed my work and life for the better.*

Prior to the onset of OCD I was, and to a large extent, still am, an optimistic, growth-oriented, and meaning-seeking person. I distinctly remember a therapy session about four months after my diagnosis. I was still in a dark place. I told my therapist Brooke that a part of what was causing me distress was that I could not see how what I was going through could possibly lead to meaning or growth. It all felt so pointless; pain with no purpose, no lesson to teach. My experience was in stark contrast to the psychology and personal-growth books I'd come across in the past, which convey to readers the importance of finding meaning even, and perhaps especially, in the depths of darkness. I assumed it was the playbook. Growth comes from struggle, right? But I could not see how OCD was related to any sort of purpose; if anything, it made me feel I had none. I shared all of this with Brooke, who herself has experienced bouts of depression. Her eyes welled up a little as she told me: *Not everything has to be meaningful and you don't have to grow from it. Why does what you are experiencing right now need to have some greater purpose? Why can't it just suck?*

---

* For those who want to learn more about OCD, I discuss my experience in further detail in my previous book, *The Practice of Groundedness*. For our purposes here, the point is that OCD is an agonizing and all-around awful experience.

Adopting a growth mindset and constructing a strong sense of meaning and purpose in one's life are incontestably healthy. These attitudes serve as a foundation for well-being and sustainable excellence in whatever it is you do—be it parenting, doctoring, writing, teaching, or starting a company. Yet there are times when life throws you such unexpected curveballs that these qualities are simply not plausible—at least not in the moment (more on this in a bit). "Such rooms in our common psychic mansion we label depression, loss, grief, addiction, anxiety, envy, shame, and the like," writes the psychotherapist James Hollis. "Such is our humanity. We are flooded by anxiety because the fact of our being out of control is no longer deniable."

During these harrowing experiences, trying to force growth, meaning, and purpose can make what you are going through worse. Not only are you hurting, scared, or grieving but you also risk becoming judgmental toward yourself about the lack of anything valuable associated with your experience. You take an overwhelmingly negative situation and inadvertently turn it into two: the awfulness of what you are going through *and* the fact that you can't even do what the self-help books tell you to. I suppose the worst offender in this category is gratitude, which I'll discuss here because it is such a stark example. No doubt, practicing gratitude is beneficial in most circumstances. Many scientific studies support it. But trying to force yourself to write down three things you are grateful for when you've just been laid off, are in a deep depression, or have recently suffered the loss of a child or partner makes little sense. I can think of few worse things to tell a depressed or grieving person than, *Why don't you reflect on all that you are grateful for right now?*

It is a catch-22. Qualities like growth, meaning, purpose, and gratitude are genuinely beneficial, and there is value in actively cultivating them. But there are also times when it is helpful to release from these notions altogether, when trying so hard to bring these qualities to fruition backfires and gets in the way. We'd all benefit from a bit more nuance.

I recently asked Brooke if she remembered the session during which she gave me permission to stop trying to find meaning and growth and, if so, what led her to that counsel. "Part of what influenced what I said in that moment was my experience trying (in vain, at times) to help other people find meaning in their painful experiences," she told me. "Sometimes it can be helpful, but other times it's not—especially if you are trying to impose or contrive it. Finding meaning and realizing growth can be a longer process, one that unfolds on unpredictable timescales."

Some cycles of order, disorder, and reorder will rapidly lead to observable growth and feelings of meaning. But this is not to be confused with a constant need to improve through every one of life's curveballs. Sometimes reorder means sitting in the wake of particularly challenging changes and gradually moving into stability without any immediate gain or discernible value. In time we tend to find meaning and grow from these events. But in the moment, being patient and gentle on ourselves is the best, and perhaps only, course to take.

Meaning and Growth Emerge on Their Own Schedule

Much like our bodies develop an immune system to fend off and heal from illness and injury, so, too, do our minds—and the two operate similarly. Let's start with a brief examination of the body and our biological immune systems. Minor injuries and ailments tend to resolve swiftly. But major ones take longer to heal, particularly if your immune system has never dealt with something like it before. You cannot trick your biological immune system: nothing that you think, say, or do will convince it that a deep gash is a minor paper cut, or that a novel coronavirus is a familiar cold. Our immune systems are extraordinary arrangements, fine-tuned over millennia of evolution. Their central role is to keep us alive and resilient, to help us move forward in the midst of unexpected biological disruptions. As such, they work as swiftly and efficiently as they can. Yet sometimes they require extended periods of time to organize an appropriate response.

The same is true for our *psychological immune systems,* a term first coined by the Harvard psychologist Dan Gilbert. Our psychological immune systems help us to filter and make sense of our lives. "If we were to experience the world exactly as it is, we'd be too depressed to get out of bed in the morning, but if we were to experience the world exactly as we want it to be, we'd be too delusional to find our slippers," writes Gilbert. When life doesn't go our way, our psychological immune systems are there to aid us in coping, healing, and moving on. In large part, they accomplish these goals by construing meaning and growth out of otherwise vexing experiences. Much like our biological immune systems, smaller and more familiar psychological setbacks yield meaning and growth more quickly than larger and unfamiliar ones. The first time a publisher turned down my writing felt like a big loss. Now when that happens, I am upset for about two minutes before getting on with my day, hopefully having learned a thing or two from the rejection. For particularly devastating and unprecedented changes—experiences like loss, illness, and identity crises—the psychological immune system does not work immediately. It takes time to marshal the resources necessary for a strong enough response. In these circumstances, premature attempts to generate a positive outlook or force meaning, purpose, and growth leave us feeling *worse* off. "These attempts are so transparent that they make us feel cheap," writes Gilbert.

Try as we might, we cannot trick our psychological immune systems with delusion any more than we can trick our biological ones. Meaning and growth emerge on their own schedule. None of this is to say there aren't certain strategies we can take to help bring about these desired outcomes. There are, and we'll examine them shortly. But as Brooke wisely counseled, we cannot force it. Trying to is counterproductive.

Why It Feels Like Difficult Times Last Forever

The period of my life that I define loosely as "being in the thick of OCD" lasted for about eight months. During that time, I struggled to have more than two decent days in a row, and more often it felt like two decent hours. It is a season of my life that seemed as if it were going to last forever. Fast-forward to today—more than six years later—and looking back, those eight months don't seem to have been such a long stretch at all. If anything, I remember them as a much smaller speck of time.

My experience is common. Research shows that when we are in the middle of difficult circumstances our perception of time slows. But when we recall those difficult circumstances with some distance, we remember them as having passed rather quickly. This distortion of time owes itself to the fact that during dark and uncertain periods, every minute tends to be occupied by a high density of distressing thoughts and feelings. It is opposite to "flow" or "peak" experiences, during which time flies because we are in the zone, hardly thinking at all. You can imagine it as the difference between watching a film frame by frame or watching it continuously. In especially challenging seasons, we experience our lives frame by frame, as a slow and decompressed progression that does not seem to be getting anywhere, let alone building up to a meaningful conclusion. But when we look back on these challenging periods, we remember them as compressed and contextualized. As such, they don't feel as devastatingly long, and we have an easier time constructing coherent and meaningful narratives. It is our psychological immune systems' method of protecting us from remembering tough periods exactly as they occurred, which would make it agonizingly hard to move on.

An extreme example is post-traumatic stress disorder (PTSD). One way it can be conceptualized is as a malfunction of the psychological immune system. Rather than process and integrate trauma into a broader narrative, effectively taking the edge off, the nervous systems of people suffering from PTSD continue to relive their terrifying episodes in vivid detail. Common symptoms include flashbacks, nightmares, and severe anxiety, as well as uncontrollable thoughts about the traumatic event. This explains why many evidence-based therapies for PTSD help people incorporate their traumatic events into a broader web of memories and other life experiences. What makes recovery so challenging is that the nervous systems of people experiencing PTSD get stuck in hyper-aroused states—which, as you're about to learn, in and of itself exacerbates the slowing of time and the stickiness of anxiety.

David Eagleman, a Baylor University professor of neuroscience, is one of the world's leading experts on the perception of time. He differentiates *brain time* from clock time, and his clever experiments show that while the latter is objective, the former is anything but. For a fascinating study, Eagleman took participants to the Zero Gravity Thrill Amusement Park in Dallas, Texas. The park, which closed during COVID, was home to perhaps the scariest ride in the world: the SCAD, short for a "suspended catch air device." Riders were lifted 150 feet into the air and positioned horizontally, so they were parallel to the ground, their backs facing down and their eyes facing toward the sky. Then they were dropped, free-falling onto a soft and netted cushion. All of the study participants who were dropped by the SCAD rated the ride as a ten out of ten on the fear scale. Immediately after they landed, Eagleman asked each participant how long their free-fall lasted. On average, participants reported it took 36 percent longer than it actually did. But when Eagleman asked participants to watch other people ride the SCAD and estimate how long those falls took, their estimates were strikingly accurate. Only when participants were in a highly aroused and anxious state—that is, during and immediately following their falls—did they feel that time slowed down.

Eagleman's work demonstrates, in part, why everything feels drawn out during particularly difficult periods of disorder. Though perhaps not as acute as dropping on the SCAD, big changes put us on high alert and in hyper-aroused states. Identifying when this is happening and being patient with ourselves is key.

There is a reason courtroom lawyers painstakingly write and refine their closing arguments. They are the last statement that jurors will hear before making a decision, which in turn means those statements will have an outsize impact on that decision. It is the same reason why if someone was recently in an argument with their significant other, even if they are a professional bowler, they will likely interpret the statement "don't cross the line" to mean don't keep prodding.

The "recency bias" says that a significant factor in how we interpret events is whatever happened last. Given that recent events hold so much power in our minds, we assume that how we feel now is also how we'll feel in the future. But these assumptions are almost always faulty because they don't account for the power of our psychological immune systems. Over time, objective stimuli—that is, what is happening right now and our associated thoughts and feelings about it—get filtered into subjective memory and woven into our personal narratives. These narratives almost always have an element of growth and meaning. If our gravest trials and tribulations didn't take on meaning, life would be too painful and we'd all be nihilists.

Combine the workings of the brain time, the recency bias, and our psychological immune systems and here is what you get: changes that trigger arousal or intense negative emotions like depression, anxiety, loss, and grief can feel utterly pointless and like they will last forever—both while we are experiencing them and shortly after. But a few days, months, and in some cases, even years down the road, we tend to reflect on these experiences with at least some degree of meaning and growth. The more difficult the change, the more time and space it generally takes. Thus, it is vital to remind yourself that even though you may feel like you are stuck in a difficult situation and the future is doomed, that rarely, if ever, is the case. Our perceptions and our ability to make accurate predictions about what will happen in the future get distorted.

In a series of studies, researchers from Harvard University (including Gilbert) and the University of Virginia set out to investigate how well people do at predicting how they'll feel about current difficulties in the future. They asked participants who were undergoing significant challenges—for example, a divorce, a layoff, or the loss of a parent—to estimate their life satisfaction, happiness, and well-being between a few months and a few years down the road. Their conclusion: "Our ability to imagine the future and foresee the transformation that events will undergo as we interrogate and explain them is limited . . . We often display an impact bias, overestimating the intensity and duration of our emotional reactions to such events."

The researchers go on to write that people "fail to anticipate how quickly they will make sense of things that happen to them in a way that speeds emotional recovery. This is especially true when predicting reactions to negative events." Anyone who has ever been dumped by a serious boyfriend or girlfriend or lost a job knows this firsthand. The first few days, weeks, and maybe even months are excruciating. But a decade down the road, most people tell themselves a story about the breakup or layoff that it was for good, or at the very least, that it wasn't so bad. In the *Tao Te Ching*, published in 400 BCE, Lao Tzu asked, "Do you have the patience to wait till your mud settles and the water is clear?" Whatever darkness you may be facing, perhaps the most important piece of knowledge to hold onto, even if just barely, is that what feels like forever now will not in the future. If this insight gives you the strength to keep showing up, then it's worth its weight in gold.

There is truth to the popular aphorism that time heals all wounds. But it's rarely just time alone. It's also what you do with it. There is a vast expanse between doing nothing and trying to prematurely force meaning and growth.

From Pain to Purpose

I first met Jay Ashman after he read my previous book, *The Practice of Groundedness*. He emailed me referencing specific parts of the book that he had found resonant. He included in his note that he'd "been through some shit." I responded with gratitude for his taking the time to read my work and reach out. A few weeks later, Jay sent me a longer note about his struggles, particularly as they relate to his identity. Again, he vaguely mentioned that he had "been through some shit," this time alluding to having "been in a gang." By then, I was deep in the process of researching and reporting for this book, so I figured it'd be worth learning a bit more about Jay and perhaps asking him if he wanted to talk. When I looked up Jay on the internet he appeared to be a total badass: shredded like the Hulk, tattoos all over his body, nose ring . . . you get the picture. I also learned that he owns a successful gym in Kansas City, Missouri. From what I could gather, his gym took a nuanced and compassionate approach to strength training. Needless to say, all of this piqued my interest. I emailed Jay, asking him if he'd be up for a chat, and he immediately replied that he was.

A few minutes into our conversation I asked him about his past, particularly his references to dark times and his involvement in a gang. He replied in vagaries. I assured him that the last thing I wanted to do was push him to share something that he wasn't comfortable sharing, but if he wanted to open up more, his words would fall on nonjudgmental ears. He paused, and then proceeded to tell me that he used to be a nationally recognized leader in the American neo-Nazi movement. I thought to myself, *Ahh, so this is the reason he was writing and speaking in such loose terms.* I quickly processed, took a deep breath, and told him, *Jay, I don't know you well but I've got more respect for you now than I did before. To work your way out of that, you must be as strong on the inside as you look on the outside.* "I don't know about that, but now that it's out of the way, let's get to talking," he said.

I learned that Jay grew up in Reading, Pennsylvania, a blue-collar town that, like so many others, suffered from the offshoring of American manufacturing. As a young child he had an auditory condition that required him to wear hearing aids, for which he was bullied incessantly. When Jay was fifteen years old, his dad died of cancer. "He literally died in my arms," Jay told me. He spiraled into depression and anger, becoming full of fury and rage. Fortunately, he channeled those qualities into football, where he excelled, eventually going on to play at Lehigh University. But upon exiting college, Jay said, "I didn't know who I was. I didn't feel like I belonged anywhere. I never liked the jocks anyways. I was really insecure and angry. I did what so many young and hurting and angry white men do: I joined the neo-Nazis." This was in 1996, when he was twenty-two years old.

Jay is a charismatic and kinetic dude. He's a big person and has a bigger personality. Unsurprisingly, he thrived in the neo-Nazi movement, rising meteorically in their ranks. However, Jay told me that he always had some cognitive dissonance. "We were taught to hate Black and Jewish people—like in a bad way. Yet I had Black friends and listened to rap music. I didn't know much about Jewish people, certainly not enough to form any sort of opinion on an individual, let alone an entire ethnicity." As much as he loved the relevance, status, and acceptance by other neo-Nazis, Jay recalls that there was always a small part of him that said, *C'mon man, is this really what you want to be doing? Is this really who you are?*

About six years into being a white supremacist, Jay was at a bar when a Black man sat down next to him. "I was wearing my neo-Nazi shirt. I wore that stuff everywhere; it was a security blanket," remembered Jay.

The man asked, "What's that on your shirt?"

"I told him straight up," Jay said. "I said it's an emblem for the neo-Nazis." The man nodded, and then they got to talking more, conversing about all sorts of topics. Over an hour later, the man got up, looked Jay in the eye, and said, "You're better than that shirt. You're better than you think you are."

By then, Jay's cognitive dissonance had been increasing to a boil. "I'd seen and done so much violence. This wasn't me," he said. That same night, Jay threw out all of his neo-Nazi shirts and got off Stormfront, the popular white supremacist internet forum. "That man saved my life," Jay said. "I will be forever grateful for his strength, kindness, courage, and compassion."

Jay, then twenty-eight, moved to New York City for the sole reason that it was a place where he could blend in and disappear. He took a job as an electrician and started personal training on the side. The same street smarts, savvy, energy, and charisma that had elevated Jay in the neo-Nazi movement also made him a great trainer, in which he found success over the next decade. "On the outside, I looked great. I was in peak physical shape, coaching high-level athletes, making money. But on the inside, I still didn't know who the fuck I was. I was still hurting," he explained.

How do you move forward with all the painful images in your head? If I am not the same person I was, then who am I? These were the sorts of questions Jay grappled with. Though he felt stuck, he kept showing up for life. He enrolled in therapy. He joined a men's group. He stayed busy. He started meditating and opening to spirituality in a way that he had always thought was woo-woo. He gave his all to his clients and to his own training, too. "I didn't feel good but I just kept showing up, one day at a time," he recalled.

In 2014, over a decade after he left the movement, Jay started to see some light. "I would have a string of good days, where I felt whole, and that was something," he recalled. He moved to Kansas City, opened a gym, and started making new friends. In 2016, Jay observed that one of America's major political parties was being hijacked by a far-right movement that was all too familiar to him. "I heard the words 'America first' in political campaigns and I knew exactly what that meant. It was the same old shit we used to say," he told me. Jay felt he needed to do something to push back, to combat the creep of white supremacy into the broader political culture. He began opening up about his past and became more politically active.

When we talked in 2022, nearly two decades after Jay left the neo-Nazis, he was doing alright. "If I can educate people now, I can do my small part to stop the spread of hate. I think this is key to my healing," he told me. "I'm finally starting to see some meaning in all the pain."

Research shows that the most common outcome of trauma is resilience and growth. This is not to negate the pain and suffering that follows trauma, nor to downplay the horrors of PTSD. It is simply a fact that most people recover and find meaning even after they have sunken down to the darkest depths. In 2010, researchers from the Medical College of Wisconsin followed 330 trauma survivors over time, many of whom required surgery at a level one trauma center. They found that as soon as six months after their traumatic event, the vast majority of survivors were already on what the researchers called a *resilience trajectory*, a path of healing and sense-making. "It is quite remarkable that such a large number of participants reported such low levels of [psychosocial distress] symptom severity," the researchers write. Interestingly, but not at all surprising given what we know about the psychological immune system, for many participants, symptoms of PTSD gradually rose, peaked at three months, and only then started to decline.

There is no one-size-fits-all trajectory to the meaning and

growth process. Physical trauma and emotional trauma are intertwined but also different. Chronic stress is different from acute stress. Injury due to assault is different from injury due to accident. It is easier to experience future meaning and growth after being laid off than it is after being raped—some horrors are truly senseless. And yet, when you look across the literature, there is a common theme. The vast majority of people eventually find meaning and grow from difficulty. The more rattling the change to someone's life, the longer it takes for that process to unfold.

What I've tried to do so far in this chapter is push against the narrative that everything has to be meaningful always. This is patently false. Rugged flexibility accepts that on occasion things feel pointless and we need to give our psychological immune systems the time they need to work effectively. Sometimes it takes days. Sometimes it takes weeks. Sometimes it takes months. Sometimes it takes years. The following sections detail a few of the most important, evidence-based tactics to work through these periods, helping us to usher in meaning and growth without forcing them prematurely. Each tactic can be applied to a wide variety of changes; and each helps with both our short-term processing and long-term thinking.

Humility and the Limits of "Fixing"

The psychotherapist James Hollis writes that what makes big and arduous changes so hard is that "the fact of our being out of control is no longer deniable." There comes a point when none of our prior strategies work. Even if we open up to and accept what is happening, expect it to be hard, conceive of ourselves fluidly, and respond not react, it doesn't mean that we'll always know what to do. When this occurs, sometimes our best bet is to surrender. This does not mean quitting on life or abandoning hope. But it does mean that we stop trying to fix, problem solve, control, or even make sense of our dire circumstances.

There is no greater source of humility, nothing that minimizes the ego more, than surrender. At first this may feel like giving up, but in the long run, it is one of the most productive actions we can take. In my own experience with OCD, it was only when I let go of any semblance of control, of any desire for meaning or personal growth, that I finally began to make genuine progress. Whatever part of me was holding on to the idea that I could somehow contrive or shape my experience was also the part of me that was holding myself back, and not just in recovering from OCD but in all likelihood in many parts of my life. (Neuroscientists might say this "part" of me is related to my brain's posterior cingulate cortex—more on this in a bit.)

When someone feels lost or broken, explains the Stanford psychiatrist Anna Lembke, they are primed for what she calls the *fundamental spiritual pivot*. "That is when we can give [our direction] over to something outside of ourselves. It can take many different forms. But the key piece is acknowledging that we are not in control, and that when we ask the universe, such as it were, to guide us or help us, that simple reorientation totally changes decision-making, it changes so many things about how we proceed in our lives." Lembke, a serious scientist who specializes in treating patients with severe substance abuse disorders and behavioral addictions, says that when her patients experience this fundamental spiritual pivot, when they throw their hands up and are forced to look to something larger than themselves for help, they start to find a way forward. "It's a total game changer, really, when you make that pivot," she explained to the podcast host Rich Roll. "And it's amazing the good things that can come from it."

Oftentimes talk of surrender is associated with a higher power or God, as is the case in traditional versions of Alcoholics Anonymous. If that works for you and your belief system, great. If it doesn't, then perhaps consider reframing "higher power" or "God" as standing in for "the universe" or "forces outside of yourself." One reason this type of surrender is so effective is that it diminishes activity in a part of the brain called the *posterior cingulate cortex*, or PCC for short. The PCC is a brain region associated with self-referential thinking, science speak for getting caught up in one's own experience. The more PCC activity that someone has, the more likely they are to get in their own way. "If we try to control a situation or our lives, we have to work hard at *doing* something to get the results we want," writes the neuroscientist Judson Brewer. "In contrast, if we can relax into an attitude that is more like a dance with [life], simply *being* with it as the situation unfolds, no striving or struggling necessary, we [can] get out of our own way."

Surrender—and the humility it spawns—not only helps us to relinquish control and cease futile efforts to prematurely force growth and meaning on chaotic situations. It also sets us up to ask for and receive help.

Asking for and Receiving Help

In his work on allostasis, the neuroscientist Peter Sterling identified a three-step process that systems undergo when confronted with significant changes. First, they try to absorb the change and adapt using "their own normal dynamic range." When that does not work, they "borrow" resources to adapt. If disorder remains persistently high, then they predict a "new normal" and gradually expand their own capacity. In other words, borrowing resources serves as a bridge to go from disorder to reorder, to stability somewhere new. When we apply this to the challenges in our own lives, it means asking for and receiving help while our psychological immune systems expand their capacity. When Jay Ashman hit bottom and felt completely lost, he sought the assistance of a therapist and joined a men's group.

Contrary to much of the popular writing on the topic, resilience is not only an inside game. Self-help is rarely enough. Studies show that asking for and receiving help is one of the most predictive characteristics for resilience. As I wrote in *The Practice of Groundedness*, the roots of massive redwood trees—towering two hundred feet above ground with trunks more than ten feet in diameter—run only six to twelve feet deep. Instead of growing downward, they grow outward, extending hundreds of feet laterally, wrapping themselves around the roots of their neighbors. When rough weather comes, it is this expansive network of closely intertwined roots that supports the trees' ability to stand strong as individuals. We are the same.

In the span of a single year, Nora McInerny lost her second pregnancy as well as her father and her husband, both of whom died of cancer. In the wake of her unimaginable suffering, she realized there is much that is broken about how society treats grief and those who are grieving. The loss of a loved one is already perhaps the worst experience anyone will ever undergo; on top of that, social norms make it feel isolating. Nora wanted to combat the shame, stigma, unrealistic expectations, and loneliness that are part and parcel of so many people's experiences of grief. Along with writing books on the topic and sharing resources on her website, McInerny started a podcast, *Terrible, Thanks for Asking*, in which people enduring severe loss and grief can share their stories. The podcast serves as a global community of people undergoing similar struggles.

"Grief is kind of one of those things, like falling in love or having a baby or watching *The Wire* on HBO, where you don't get it until you get it, until you do it," McInerny explains in her 2018 TED Talk on the subject, which has been viewed over six million times. She goes on to explain that no one should have to walk this path alone, a principle that has guided all of her work.

McInerny's words remind me of how I think about depression. Being depressed is like being on one side of a river that looks exactly the same as the other side but *feels* very different. People on the other side tell you to *brighten up,* that *you'll be okay,* and *not to worry, everyone gets sad from time to time!* But none of that is really helpful. What is helpful is when someone who has spent time on your side of the river jumps across and joins you. You may think, *What the hell is that person doing? Why are they coming over here into this place with me?!* And then they tell you, *I came over here because I have been in these waters before and I know how awful they are.* They proceed to take your hand and, if possible, help you wade your way out.

Whether you are devastated because of an acute loss, or merely saddened because of the state of the world or because your athletic event or big presentation went poorly, asking for and receiving help empowers you to hold on to the knowledge that what you are feeling is real *and* if you just keep showing up you'll be able to move forward, even if that feels impossible now. Perhaps nothing brings people together more than shared hardship. A central part of the reason our species evolved to live in groups is because it is near impossible to get by otherwise. Pain and suffering are never easy, but they are a bit less hard when they are held together.

Voluntary Simplicity

For a series of studies published in the prestigious journal *Nature*, the University of Virginia interdisciplinary scholar Leidy Klotz and colleagues presented participants with a set of problems across a broad range of topics. These included design schemes, essays, recipes, travel itineraries, architecture, and even malfunctioning miniature golf holes. They then asked participants to make changes to improve each. What they found is that the vast majority of people tend to overlook the option to subtract parts. Instead, participants immediately assume the best way forward is to add, even when subtracting is clearly a better option. When I asked Klotz about why this is the case, he told me that "such a large part of our culture is to have more, do more, and be more, so people kind of assume more is always the answer, but it's not."

Klotz's work reminds me of what the meditation teacher Jon Kabat-Zinn calls *voluntary simplicity*, or intentionally choosing to simplify our lives by taking out clutter, be it physical, psychological, or social. We don't realize how much of our daily stress is a consequence of having too much to do and track, most of which is not valuable, let alone essential. Particularly when the world around us feels big, confusing, and overwhelming, it can be helpful to scale back, going small and minimal. This does not mean that we ought to shut down completely or sequester and close ourselves off. Rather, we ought to focus on what matters most, what gives us a chance at feeling good and persisting, and then be okay subtracting everything (or at least everything we can) that does not. Two of the best ways to introduce voluntary simplicity are through routines and rituals.

Routines serve as bedrocks of predictability, creating a sense of order amidst disorder. They are also helpful because they automate action, simplifying life by allowing you to show up and get going without having to exert any additional energy psyching yourself up or thinking about what you ought to do. As we discussed in the previous chapter, even the smallest victories—writing one sentence, going out for a short run, knitting a single square of a quilt, doing a load of laundry—release the neurochemical dopamine, which fuels our drive to keep going in whatever it is we are doing, and also in life itself.

Research shows that the same brain region that is activated with cocaine (the striatum) is also activated with accomplishment. This likely explains why so many people numb their pain by throwing themselves into work, sometimes becoming workaholics. Maybe this is not ideal, but who is to say? Assuming the work is meaningful, there are plenty of worse outlets. Another example is sport. For instance, many ultramarathon runners are in recovery. Maybe they've traded one addiction for another, but long-distance running tends to be much healthier than using illicit substances. Perhaps the best way to think about this trade-off is that throwing yourself into work (or other activities) is a good strategy to help you keep going amidst life's most severe challenges, but you don't want to become dependent on it as a long-term analgesic. Even simpler, we can ask if our throwing ourselves fully into an activity is helping or hurting us, if it is expanding our lives or contracting them. What may start out as the former can turn into the latter.

Closely related to routines are rituals, specific activities that people perform at regular intervals, during periods of stability and change alike. "[Rituals] open a space in which to host thoughts that I would otherwise find silly or ridiculous: a voiceless awe at the passing of time. The way everything changes. The way everything stays the same," explains the writer Katherine May. Examples include weekly religious services, monthly neighborhood dinners, lighting candles every morning, going on a group bike ride every Sunday, and so on. Much like routines, rituals provide structure and stability when everything around us is changing. They also serve as a reliable source of voluntary simplicity: the world may be chaotic but I know that every Friday morning I go for a long walk in the woods with my dog, and in that time and space life feels simpler and more manageable.

In his book outlining allostasis, *What Is Health?*, Peter Sterling writes about what he calls *sacred practices,* "where 'sacred' means 'reverence for the ineffable'—what casual speech cannot express." Examples include ceremonies of singing, dancing, exercising, praying, listening to music, and so on. "The circuits that produce and process these activities occupy substantial cortical [brain] territory," he writes. "The neural investments to produce and process music, art, drama, and humor indicates their central importance to our success." In layperson's terms, if sacred practices were not inherently advantageous for our survival, evolution would have programmed these precious brain circuits for something else. Yet here we are, well equipped. Perhaps this is because during periods of immense disorder, when we risk feeling completely unmoored, rituals help keep us grounded, providing some sense of stability in otherwise uncertain times. As such, they are vital to our persistence.

Real Fatigue Versus Fake Fatigue

One of my coaching clients, whom I'll call Melanie, is a thirty-nine-year-old entrepreneur. She had been undergoing a series of changes and struggling with fatigue—nothing too severe, but a general sense of exhaustion or, in her words, "not feeling as sharp and energetic as I'd like." The first solution that came to mind was simple: rest. But she had been resting for over a month, scaling down her work and personal obligations, yet she still felt sluggish.

Melanie's situation is common. It illustrates what I've come to think of as the difference between two types of fatigue: when your mind-body system is truly tired, or what I call *real fatigue*; and when your mind-body system is tricking you into feeling tired because you are stuck in a rut, or what I call *fake fatigue*. It is important to differentiate between these two sensations since the response that each requires could not be more different. Real fatigue calls for shutting things down and resting. Fake fatigue calls for not taking the feeling of exhaustion too seriously but rather working your way out, nudging yourself in the direction of action, and making a commitment to show up and get started on what you had planned.

It is easier to discern between real fatigue and fake fatigue when you are dealing predominantly with your body. Here, feedback tends to be more objective—your muscles become sore, your heart rate increases, or the speed at which you walk or run declines. For more generalized and predominantly psychological fatigue, however, clear metrics are lacking. This means that you've got to feel your way into the right response. Sometimes this means staying in bed or on the couch; other times it means forcing yourself to get going.

Generally speaking, the cost of pushing through real fatigue is greater than the cost of acquiescing to fake fatigue. Going too hard for too long and repeatedly pushing yourself over the edge results in burnout, which research shows can take months—and in severe cases, years—to reverse. Perhaps the safest bet, then, is to treat the onset of exhaustion as if it were real fatigue. Take a day off, or a few. Sleep a little extra. Disconnect from digital devices. Spend time in nature if you can. Reexamine your routine, and if something seems haywire, make adjustments. If you do all of this and you still feel malaise, then it is probably worth exploring what happens if you nudge yourself into action.

A common example of fake fatigue is the exhaustion that accompanies big life changes such as loss, grief, changing jobs, relocation, or retirement. Your brain is doing everything it can to trick you into staying in bed all day when the best thing you can do to feel better might be to get up and go, to engage in the sort of behavioral activation that we learned about in the prior chapter. This isn't to say that the sensations of lethargy and dullness are not real—they are, and they can be quite paralyzing. But those sensations, as far as we know, are usually not solely organic—not caused by a lack of sleep, an expenditure of physiological resources, or something wrong in the body. If they were, taking action would make the situation worse. But, as the research shows, behavioral activation tends to make these situations better, especially when it is supported by help-seeking and community.

Fake fatigue is common not only following major disruptions but also on a smaller scale. For instance, when it came time for me to shift focus from promoting my previous book to writing this one, I kept putting it off. Though it was by no means a massive or arduous change, it was a change nonetheless. Every day that I had scheduled to start writing I felt tired! So I rested. And rested some more. After about three weeks of this, I decided to take some of my own medicine and force myself to just get going, whether I felt like it or not (to be sure, I did not). Three days later, I was in a writing groove that lasted for over a month. More rest would have only deepened my rut. I needed to work my way out.

There is one additional layer of nuance here, and it is an important one. Sometimes breaking out of chronic exhaustion and malaise requires combining both of the above strategies. You may be experiencing real fatigue and thus need rest. After a week of rest, your mind-body system may be recovered but now latched onto the inertia of doing nothing. At this point, the strategy shifts to behavioral activation. In sports, this is why tapers (the prolonged period of rest before a big event) usually end with a few short, intense efforts, which serve to wake the body up and snap it back into action. I suspect our minds operate in the same way. After we experience a substantial change, we may very well require a longer period of downtime. And that extended downtime works great—until it becomes the very thing that gets in our way.

Where does this leave us? Our best bet is probably to think of managing fatigue as an ongoing practice. If you pay close attention to how you feel, what you do in response, and what you get out of it, over time you'll become better at differentiating between real and fake fatigue. The first and most important step is realizing that not all sensations of fatigue mean the same thing. For those accustomed to always pushing through exhaustion, maybe you need a bit more rest. For those accustomed to always resting, perhaps you'd benefit from a bit more pushing, from more of a "mood follows action"* mentality. There is a time and a place for each.

Flowers Grow from Mud

In perhaps my favorite book ever, *Zen and the Art of Motorcycle Maintenance,* the middle-aged narrator and his young son, Chris, are on a cross-country motorcycle trip. When they reach the

---

* An axiom I first heard from my friend Rich Roll.

mountains of Montana, they hear a rockslide and Chris asks his dad why these happen.

"It's part of the wearing down of mountains," explains the narrator.

"I didn't know mountains wore out," responds Chris.

"Not wore out. Wore *down*," says the narrator. "They get rounded and gentle... Mountains look so permanent and peaceful, but they're changing all the time and the changes aren't always so peaceful."

I suspect the narrator, who at the time was going through his own identity crisis, wasn't talking just about mountains but also about himself, and about all of us. Nobody escapes life unscathed. The biggest and harshest changes in our lives are akin to rough weather on a mountain. They wear down our edges and make us softer and gentler. The result is that we gain compassion, both for ourselves and for others.

Conventional wisdom on getting through challenges says that on one extreme there is taking responsibility and picking oneself up by the bootstraps, and on the other there is taking it easy and showing oneself boundless love. While these are often pitted against each other, the truth is that they are complementary: In most circumstances you need at least some measure of both. Yet again, non-dual thinking, the unsung hero of this book, makes another appearance. The best approach is to combine fierce self-discipline *and* fierce self-compassion. Regularly practicing self-compassion makes you fearless. If you know that you can be kind to yourself, then you can go to tough places, knowing that you've got your own back. Showing up during periods of disorder can be hard, no doubt. But with self-compassion, you make it just a bit easier. Being kind to yourself in the midst of struggle and hardship affords you the resilience you need to endure, persist, and flourish.

Self-compassion is not automatic; like any other important quality, it must be developed. Notice when you are being particularly hard on yourself. How does it make you feel? What would it look like to change the self-talk? This is not about brushing off every misstep; it's about not wasting energy beating yourself up. When you enter a ruminative or self-judgmental spiral, ask yourself: *What would I say to a friend in this situation?* We tend to be much kinder and wiser when we are giving advice to our friends than when we are giving advice to ourselves. You can also call upon a mantra, which snaps you out of your head and puts you back into the present moment. One that I use all the time, with both myself and my coaching clients, is simple: *This is what is happening right now; I'm doing the best I can.* Another benefit of this particular mantra is that if it isn't true—if I am not doing the best I can—then I realize that too, and kindly give myself the chance to do better.

When we find ourselves in Dante's dark wood with no clear path out, when we experience surrender and the radical humility it births, when we ask for and receive help, when we struggle to show up and stick with even the barest routine—hopefully as a result we become at least a little kinder toward other people who are suffering too. What comes around goes around. The same people we depend on for comfort and solace in our periods of disorder will eventually depend on us during theirs. What good is it to suffer if we don't use it to bring ourselves closer to others? If we don't harness our shared impermanence, and all of the pain and difficulty that occasionally come with it, toward weaving the relational safety net that, over time, is tasked with supporting us all? In these fast-paced times of "optimization" and "efficiency," we'd be foolish not to slow down and ensure that we are also doing the essential work of nurturing intimate relationships and community. When the going gets tough, arguably nothing is more important.

Thich Nhat Hanh, or Thay, as his students reverentially called him, famously taught "no mud, no lotus." A lotus is an exquisite species of flower. Its colors are bright and striking, its petals open and inviting. What makes lotus flowers so fascinating is that they grow out of mud. Thay taught that suffering is like mud. But we can transform suffering into a beautiful and radiant lotus flower: compassion. It's a transformation that does not usually happen when you are in the thick of a trying experience. But if you can keep showing up, over and over and over again, eventually you'll get to the other side. Once you do, odds are you'll have gained a fair share of compassion along the way. With each significant cycle of order, disorder, and reorder we endure, we become a bit kinder and softer toward ourselves and a bit kinder and softer toward others. If anything good comes of suffering, it is this.

Hard Times Are Always Hard—But with Practice They Get Easier

In a multiyear study of more than 2,000 adults aged 18 to 101 published in the *Journal of Personality and Social Psychology*, University of Buffalo psychologist Mark Seery and colleagues found that people who had experienced medium levels of adversity were both higher-functioning and more satisfied with their lives than those who had experienced extremely high levels of adversity as well as those who had experienced hardly any adversity at all. Equally important, the people who experienced mid-level adversity also coped better with future challenges, leading the study's authors to conclude, in an evidence-based remix of Nietzsche, "in moderation, whatever does not kill us may indeed make us stronger." It is an important study because it shows that we get better at navigating disorder over time. And yet, it also shows that extreme varieties of disorder—such as rape, assault, murder, and war, for example—are never desirable.

I feel strongly that we should not glorify or romanticize any kind of suffering. Suffering sucks. Period. But it is also part of the human experience, a nonnegotiable consequence of living and caring and loving in an impermanent world. Loss, grief, and sadness are the price we pay for love, care, meaning, and joy. To be rugged and flexible is to hold space for and endure both. Here's the poet, Mary Oliver:

> We shake with joy, we shake with grief.
> What a time they have, these two,
> housed as they are in the same body.

For each period of disorder we experience, we get a little better at navigating future ones. The next time a massive change shakes us to our cores we may still feel terrible at first, but a part of us, perhaps just 1 percent more each time, knows that a benefit of impermanence is that it does not discriminate: The lows pass too and in all likelihood, we'll derive at least some meaning and growth from our experience, even if it takes time. "There is enlargement of the spirit purchased by suffering and humility, but enlargement nonetheless. We may like it less at first, but we will be larger for going there," writes James Hollis. We've got to try as hard as we can to believe Hollis's words and hold on to them when things fall apart. If we can get to the other side of our most challenging trials and tribulations, strength, meaning, growth, kindness, and compassion await.

Making Meaning and Moving Forward

- Growth and meaning unfold on their own schedule; we need to give our psychological immune systems time to process the significant changes and disruptions in our lives.
- Our perception of time slows during difficulties; just knowing this helps us to be patient and persist; what feels awful today almost certainly won't feel as bad in far-off tomorrows.
- Though we cannot force meaning and growth, a few concrete tactics can help us to usher them in:
    → Practice humility and surrender, which does not mean doing nothing but rather releasing from the need to fix or control unfixable or uncontrollable situations.
    → Ask for and receive help: beware of getting pulled into the vortex of extreme optimization and productivity at the expense of nurturing friendships and building community.
    → Practice voluntary simplicity, develop routines, and create rituals.
    → Separate real fatigue from fake fatigue—remember that the former calls for rest and the latter calls for nudging yourself into action.
    → Do what you can to let your suffering turn into compassion for yourself and others.
- With each major cycle of order, disorder, and reorder that we navigate, the next one becomes just a little easier.

Conclusion: Five Questions and Ten Tools for Embracing Change and Developing Rugged Flexibility

Our biggest personal and collective challenges revolve around change. For individuals these include aging, illness, gain, and loss. For organizations it means changes in where people work, how people work, and why people want to work in the first place. Societally it means climate change, demographic change, and geopolitical change. Peter Sterling, the University of Pennsylvania professor who developed the concept of allostasis, defines *health* as the "capacity for adaptive variation." Disease, he writes, "is the shrinkage of this capacity." A big part of the reason that our culture is so unhealthy and at such *dis*-ease—physically, emotionally, intellectually, socially, and spiritually—is because we lack the requisite skills to navigate change. My hope is that this book serves as a much-needed corrective and sheds light on those skills.

The stakes are too high to continue as we have. It is unhelpful to rotely resist change, but it is also unhelpful to thoughtlessly go along with it, like automatons at the whims of forces larger than ourselves. If we take either of these two prevailing approaches, we'll continue to exacerbate an obesity crisis (changing food supply), an attention crisis (changing technology), a loneliness crisis (changing social norms), a democracy crisis (changing politics), an environmental crisis (changing climate), and a mental health crisis, resulting in large part from a combination of all the above. If we are to reclaim our health, let alone have any chance at flourishing, then we need to transform our relationship with change by becoming more active participants, understanding that we can shape change as much as change can shape us.

In the middle of 2022, as I was finishing my first draft of this book, Rebecca Solnit penned "Why Did We Stop Believing That People Can Change," a beautiful essay arguing that, now more than ever, we need to be conversant with change. "[Our] belief in the fixity rather than the fluidity of human nature shows up everywhere," and to great detriment, she writes. We lock ourselves into who we currently are or who we once were, and we do the same to others and to the world as a whole. The result is that growth and progress, and the basic hope upon which growth and progress rely, are stifled. Solnit goes on to highlight our perilous overreliance on this-or-that thinking. "Perhaps some of the problem is [our] passion for categorical thinking, or rather for categories as an alternative to thinking." Though she doesn't explicitly mention it, the best antidote is the sort of non-dual thinking that we've discussed at length in this book. Solnit ends her essay by calling for "a recognition that people change, and that most of us have and will, and that much of that is because in this transformative era, we are all being carried along on a river of change."

I wholeheartedly agree. It is why—in the midst of the COVID pandemic, the backsliding of Western democracies, the transformation of the workplace, and war on the European continent; in the midst of watching my little boy grow up so quickly, having a new baby girl, moving across the country, experiencing the great high of professional success as a writer and the painful low of becoming estranged from some of my closest family members; in the midst of hearing from friends, clients, colleagues, and neighbors about all the overwhelming changes in their lives—I wrote this book. A decade from now the core objects of change may be different, but there will be change nonetheless. I believe that rugged flexibility and the qualities that underlie it can help us to skillfully navigate our own respective cycles of order, disorder, reorder, which is to say our lives, and to become better community members too.

Five Questions for Embracing Change

To solidify and make concrete what we've learned, it can be helpful to ask yourself the following five questions. Language is a powerful tool. Once you put words to an unnamed thought, feeling, or concept, you shine a light on it and make it tangible—and thus you can wrestle with it in new and meaningful ways. Even if you don't have immediate answers, simply asking these questions will help you weave rugged flexibility into the fabric of your life.

1. **Where in your life are you pursuing fixity where it might be beneficial to open yourself to the possibility— or in some cases, the inevitability—of change?**
   The Zen master Shunryu Suzuki, who helped to popularize Eastern philosophy in America in the early 1960s, was known to say just two words could summarize all his teachings: "Everything changes." In physics, it's the second law of thermodynamics: as time goes on, the net entropy, the degree of change and disorder in living systems, always increases.

Our resistance to this basic and easily observable fact causes unnecessary suffering. There is no denying that in and of itself change can be painful. But we make it worse by clinging and desperately wanting certain things to stay the same when that's impossible. Think back to our math lesson from chapter 2: suffering equals pain times *resistance*.

Suzuki also taught that if you look deeply into any phenomena, eventually you'll see its truth. Pay close attention to the areas of your life where you feel tension, and odds are you'll discover at least some resistance to change. We've discussed common offenders at length including aging, relationships, big projects at work, external measures of success, plans for your future, and episodes from your past. When you identify specific areas of resistance, explore what it would look like to loosen your grip, even if only a little.

Our progress depends upon nurturing the parts of ourselves that accept, work with, and integrate change so that they are more powerful than the parts of ourselves that stubbornly, and sometimes even dangerously, resist it. Striving for fixity is an enormous weight. See if you can drop it.

2. **In what parts of your life are you holding on to unrealistic expectations?**
   As we learned, our happiness is a function of reality minus expectations. There's a good argument that the best definition of *reality* is "change." It follows that if we expect things will never change, well, we'll spend a lot of our lives unhappy due to wildly faulty expectations.

   Where in your life are you wearing rose-tinted glasses? How could you take a more accurate view? What would it look like for you to accept the world on its terms without giving up hope that it can become better?

There is a story of a wise Thai Forest elder named Achaan Chaa who held up his favorite glass in front of his students and said, "You see this goblet? For me this glass is already broken. I enjoy it; I drink out of it. It holds my water admirably, sometimes even reflecting the sun in beautiful patterns. If I should tap it, it has a lovely ring to it. But when I put this glass on the shelf and the wind knocks it over or my elbow brushes it off the table and it falls to the ground and shatters, I say, 'Of course.' When I understand that the glass is already broken, every moment with it is precious." Chaa's example is a lofty aspiration, no doubt, but one worth keeping in mind.

3. **Are there elements of your identity to which you cling too tightly?**
We all wear many hats. A few examples include parent, partner, child, sibling, writer, employee, executive, physician, friend, neighbor, athlete, baker, artist, creative, attorney, and entrepreneur. Take an inventory of your own identities. Are there any upon which you are over-reliant for meaning and self-worth? What would it look like to diversify your sense of self? Even if you desire to go "all in" on a certain endeavor, how might you ensure that you don't leave others completely behind? It is okay to put all your eggs in one basket, so long as you have other baskets available when the one into which you are currently pouring yourself changes.

Even better is to challenge yourself to integrate the various elements of your identity into a cohesive whole. This allows you to emphasize and deemphasize certain parts of your identity at different periods of time. In my own life, there are times when I lean heavily into each of my main identities—father, husband, writer, coach, friend, athlete, and neighbor. I've learned the hard way that when I minimize any of these identities too much, things tend not to go well. But when I focus on keeping all of these identities strong, it ensures that when things falter in one area of my life I can rely on the others to energize and pick me up, which usually helps me to stay grounded and navigate whatever challenge I am facing.

4. **How might you use your core values—the rugged and flexible boundaries of your identity—to help you navigate the challenges in your life?**

Your core values represent your fundamental beliefs and guiding principles. They are the attributes and qualities that matter to you most. It is helpful to come up with three to five (an extensive list of example core values is in the appendix on page 202). Define each in concrete terms, and then consider a few ways you can practice it. If you are struggling to come up with core values, think of someone you look up to and respect. What is it about that person you admire? You can also imagine an older and wiser version of yourself looking back on current you. What characteristics would make older, wiser you proud?

When faced with change, disruption, or uncertainty, ask yourself what it might look like to move in the direction of your core values. At the very least, how might you protect them? The manner in which you practice your core values will almost certainly change—being able to manifest them in new ways and situations is the crux of flexibility. And though it isn't necessary, it is also normal for your core values to shift over time. Navigating the world using your current core values is what guides you to your new ones. Your core values are a driving force in your personal evolution; they are the chain that links where (and who) you are to where (and who) you'll be.

Ruggedness without flexibility is rigidity, and flexibility without ruggedness is instability. Consider where you fall on this spectrum, and what a healthy middle ground might look like. If you are too flexible, challenge yourself to more firmly uphold and practice your values. If you are too rugged, challenge yourself to broaden the ways in which you apply them.

5. **In what circumstances do you tend to react when you would benefit from responding, and what conditions predispose you to that?**

    Reacting is rash, automatic, and careless. It places you in autopilot mode. Responding is calculated and deliberate. Many people, myself included, tend to fall into somewhat predictable patterns of reacting in specific situations. Perhaps it is when you are engaging with a certain colleague or family member. Or maybe it is when a particular topic of discussion arises. Or whenever you are presented with bad news. Once you identify these situations, you can bring additional awareness to them, challenging yourself to slow down and respond instead.

    It is also worth considering what conditions predispose you to reactivity more broadly. Do you carry a shorter fuse when you are spending too much time on social media? When you are watching certain kinds of television programming? When you have too much on your plate and you feel like there is no open time or space in your day? Once you identify these triggers you can work to eliminate, or at the very least, minimize them from your life.

Ten Tools for Developing Rugged Flexibility

So long as we are alive, we'll find ourselves in ongoing cycles of order, disorder, and reorder. Adeptly navigating these cycles demands rugged flexibility. To be rugged is to be tough, determined, and durable. To be flexible is to respond to altered circumstances or conditions, to adapt and bend easily without breaking. Put them together and the result is a gritty endurance, an anti-fragility that not only withstands change, but also thrives in its midst. Here are ten of the most important ways to practice rugged flexibility in your daily life.

1. **Embrace non-dual thinking.**

    While some things in life truly are either/or, many are both/and. Philosophers call this kind of thinking *non-dual*. It recognizes that the world is complex, much is nuanced, and truth is often found in paradox and contradiction: not this *or* that, but this *and* that. Non-dual thinking is an important, albeit spectacularly misunderstood and underused concept in many facets of life, including when it comes to change.

    One way to differentiate knowledge from wisdom is that knowledge is knowing something and wisdom is knowing when and how to use it. Inherent to non-dual thinking is realizing that many concepts and tools work great until they get in your way. For instance, the goal of rugged flexibility is not to be stable and therefore never change. Nor is the goal to sacrifice all sense of stability, passively surrendering yourself to the whims of life. Rather, the goal is to marry these qualities, understanding when and how to stand firm and when and how to adapt. As the Nobel Prize–winning psychologist Danny Kahneman used to tell his students, "When someone says something, don't ask yourself if it is true. Ask yourself what it might be true of." Another helpful question to ask yourself: *Is this view or approach helping me right now?* If the answer is yes, keep using it. If the answer is no, then shift, all the while realizing that how you answer will probably evolve over time, and that's okay.

2. **Adopt a *being* orientation.**
   A *having* orientation means that you define yourself by what you have; thus, you are inherently fragile since those objects, identities, and pursuits can be taken away. Anything that you so desperately want to own inevitably ends up owning you. A *being* orientation, on the other hand, means you identify with the deepest and most enduring parts of yourself: your core values and your ability to respond to circumstances, whatever they may be. A being orientation is dynamic and thus advantageous for working with change. If you find yourself becoming overly attached to any one person, place, concept, or thing, broaden the story you tell yourself about yourself. Instead of thinking, *I am the person who has X, Y, and Z,* try thinking, *I am the person who does X, Y, and Z.*

3. **Frequently update your expectations to match reality.**
   The human brain functions like a prediction machine that is constantly trying to anticipate reality. You feel (and do) best when your reality is aligned with, or perhaps slightly better than, your expectations. Try to set appropriate expectations, and when unsure, err on the side of being cautious and conservative. When an unforeseen change occurs, do everything you can to see it for what it is and update your expectations accordingly. The longer you cling to old expectations, the worse you'll feel and the more time and energy you'll waste when you could be working on what is happening in front of you instead. *Here is what I was hoping for or thought would happen. Here is what is actually happening. Since I live not only in my own head but also in reality, I need to focus on the latter.*

4. **Practice tragic optimism, commit to wise hope, and take wise action.**

In an interview with *The Atlantic* shortly after his album *Letter to You* was released, a seventy-one-year-old Bruce Springsteen suggested that the heart of wisdom is learning "to accept the world on its terms without giving up the belief that you can change the world. That's a successful adulthood—the maturation of your thought process and very soul to the point where you understand the limits of life, without giving up on its possibilities."

Gently push yourself to follow Springsteen's poignant counsel; to acknowledge, accept, and expect that things are going to be hard, that sometimes impermanence hurts. Then, do what you can to trudge forward with a positive attitude nonetheless. Called *tragic optimism,* this is a profound conduit to developing compassion and connection. As far as scientists know, the human species is the only one that can look ahead and understand that everything, including what we love, is going to change. Impermanence is a shared vulnerability and thus it can bring us together. This communion not only helps us to keep going, but it is also one of the best parts of being alive. As the Harvard psychologist turned spiritual teacher Ram Dass used to say, "We are all just walking each other home."

If tragic optimism is a mindset, then wise hope and wise action are its concrete consequences. Committing to wise hope and taking wise action means not wallowing in despair but not becoming a Pollyanna either. It's about doing something productive instead. Wise hope and wise action require accepting and seeing a situation clearly for what it is, and then, with the hopeful attitude necessary, saying: *Well, this is what is happening now, so I will focus on what I can control and do the best I can. I've faced other challenges and other seasons of doubt and despair, and I've come out the other side.* Remember that hope is most important in the situations when it is hardest to hold on to it.

5. **Actively differentiate and integrate your sense of self.**
   Complexity is crucial for persisting through periods of change and disorder. It requires both differentiation and integration. Differentiation is the degree to which you are composed of parts that are distinct in structure or function from one another. Integration refers to the degree to which those distinct parts communicate and enhance each other's goals to create a cohesive whole. Think about the distinct elements in your own life and how they work together. If you are not differentiated enough, how might you become more so? What pursuits could you start, maintain, or spend more time with? The same goes for integration. How could you mold the distinct parts of your identity into a cohesive narrative?

6. **View the world with independent and interdependent lenses.**
   People tend to adopt one of two selves in relation to the various roles and environments they inhabit. An independent self views itself as individual, unique, influencing others and its environments, and free from constraints. An interdependent self views itself as relational, similar to others, adjusting to situations, and rooted in traditions and obligations. Once you are aware of these lenses, you can choose when to wear each. Consider how and when you might switch between an independent and interdependent lens in your own life. The former is advantageous when you want to make something happen and have a high degree of control. The latter is advantageous when

you are in a more chaotic environment. When you start on a big project, you can take stock of which lens you suspect will benefit you most. If you find yourself butting up against a brick wall, become aware of which lens you are using, and see if approaching the situation from the other might help. Remember to think non-dually. Even in a single project, there will likely be moments that benefit from deploying each lens.

7. **Respond to change with the 4Ps.**
Skillfully responding to change requires creating space between an event and what you do, or don't do, about it. In that space, a *pause*, you give immediate emotions room to breathe and thus you come to better understand what is happening—that is, you *process*. As a result, you can reflect and strategize using the most evolved and uniquely human parts of your brain to make a *plan*, and only then *proceed* accordingly. To help you pause, label your emotions. To help you process and plan, try one of the self-distancing techniques: giving advice to a friend, practicing mindfulness meditation, or experiencing awe. The biggest barriers to proceeding are self-doubt and paralysis by analysis. The best way to overcome them is to treat your first moves as experiments. Lower the bar from needing to take the absolutely right or perfect action to trying something new and learning from it. If hindsight proves your actions useful, keep going down the same path. If hindsight proves them unsuitable, adjust course, perhaps repeating the first 3Ps—pause, process, and plan—before proceeding again.

8. **Lean on routines (and rituals) to provide stability during periods of disorder.**
Routines offer a sense of predictability and stability when everything around you is changing. They also help you to

activate by automating decisions so you need not rely as much on willpower and motivation, both of which tend to be in short supply during significant difficulties. But here's the catch: although routines can be magical, there is no magic routine. What works for one person might not work for others. The best way to develop an optimal routine is through astute self-awareness and experimentation. Pay attention to what you do and what you get out of it. It is beneficial to develop routines, and their close cousin, rituals, during times of relative stability. This way, they'll be well worn and easier to call upon when chaos strikes.

There are, of course, certain behaviors that are near universally effective, such as exercise, sleep, and social engagement. But even then, there is no optimal time, place, or way to engage in these behaviors. You've got to figure out what works for you. There is also a danger in becoming overly attached to your routine. If for whatever reason you can't stick to it—you're traveling, your special coffee shop closes, whatever elixir you order from your favorite podcast's advertising goes out of business—you won't know what to do. It's like a Zen koan: the first rule of routines is to develop one and stick with it; the second rule is to be okay with releasing from it.

9. **Use behavioral activation.**
Sometimes when we are stuck and feeling exhausted—emotionally, physically, socially, or spiritually—the best thing we can do is rest. But at a certain point, rest creates inertia. Our minds and our bodies are as recovered as they are going to be. Yet we still feel off. At this point, we can likely benefit from deploying a psychological concept called *behavioral activation*. First developed in the 1970s by the clinical psychologist Peter Lewinsohn as a way to help people work through depression,

apathy, and other entrenched negative states of mind, behavioral activation is based on the idea that action can create motivation, especially when we are stuck or in a rut.

To be clear, this is not about trying to think positive thoughts, a mantra that became a pillar of the self-esteem movement in the last century, with mega-best-selling books such as 1952's *The Power of Positive Thinking* arguing—we now know, falsely—that if you just think positive thoughts and suppress negative ones, you'll gain health, wealth, and happiness. If anything, research has shown that those strategies often backfire: The more you mentally try to change how you feel, the more stuck in your current mood you're likely to end up. You simply cannot think or will yourself to a new state of being.

The challenge with behavioral activation is mustering enough energy to start acting on the things that matter to you. When you feel down, unmotivated, or apathetic, give yourself permission to feel those feelings but not dwell on them or take them as destiny. Instead, shift the focus to getting started with what you have planned in front of you, taking your feelings, whatever they may be, along for the ride. Doing so gives you the best chance at improving your mood. It can be helpful to think of this initial oomph as activation energy. Sometimes we need more, and sometimes we need less. When we are in a rut, even the little things require more, and that's okay. It may take some extra work to overcome the initial stasis and friction. But the laws of physics apply to our minds too: the more we get going, the easier it becomes.

10. **Don't force meaning and growth; let them come on their own time.**

Research shows that most people grow and find meaning in even the most harrowing struggles. But the bigger the difficulty, the longer this process takes, and it cannot be forced. Trying to prematurely impose meaning and growth on yourself (or an experience) almost always backfires. You all too easily end up taking a negative—for example, the loss of a job, the loss of a loved one, or a traumatic injury—and turning it into a double negative: the awfulness of what you are going through *and* the fact that you can't even do what the self-help books tell you to.

During life's most significant challenges, the stuff you can't possibly imagine until you find yourself in the thick of it, give your psychological immune system the time and space it needs to marshal an appropriate response. There is no need to place extra pressure on yourself. Simply showing up and getting through is enough. You won't be the same and not everything will necessarily be okay, but there is a high probability that you'll find at least some meaning and growth, even if that seems impossible when you are in the middle of the hardship itself. Be kind and patient with yourself—hard as it may be—and do what you can to lean on others for support. We are all in this together.

Acknowledgments

This book, along with any of my others, does not happen without Caitlin. Period. She is the best partner for me. I am grateful for her every day. My son, Theo, made this process so much more fun than my prior books. It is hard to take yourself too seriously with a five-year-old asking one hundred thousand (truly!) questions about the publishing process. My daughter Lila was new on the scene. If enough readers made it this far, that means you'll probably get to help me on my next book, Lila! Also, if I am doing this with integrity, which I am, thanks are in order to my roommate, Sunny (feline); snuggle buddy, Bryant (also feline); and my best friend, Ananda (canine). Y'all force me to get up from my chair multiple times every writing day, which is (mostly) a good thing.

My core team: Steve Magness for being my *other* partner (and Hillary, for not getting too upset about Steve's and my phone time). Laurie Abkemeier, my agent, publishing coach, and first-line editor, for being the best guide on this ever-changing publishing path. Chris Douglas, for helping take what Steve and I do to the next level and running *The Growth Equation* with the intentionality and thoughtfulness it deserves.

This book had such amazing support and input. Thank you, Courtney Kelly, for helping research many of the wonderful character-driven stories you just read. Mara Gay for multiple early reads and rounds of feedback (and also for being such a wonderful longtime friend!). Tony Ubertaccio for helping me dial in the introduction to this book, and for many wonderful long walks in the woods. My author mastermind group (and some of my best buddies): Dave Epstein, Cal Newport, Adam Alter, and Steve Magness. My mentors Mike Joyner and Bob Kocher. My best friend, Justin. My "spiritual friendship" friend Brooke. My brother, Eric. All my wonderful neighbors—writing is so much easier when you live in a great community! And Zach, thanks for keeping my body healthy during the time I wasn't glued to my writing chair; it's truly amazing how much I can deadlift from a place of love and not fear, and I owe that to you.

I also feel enormous gratitude toward my editor, Anna Paustenbach, and the entire team at HarperOne. I am especially thankful to have found someone who immediately saw the value of this book and its central message, no explanation needed. Working with Anna has been like working with a sister, only one you don't ever fight with and from whom you learn a ton! One lesson in particular from Anna that I'll carry in my writing tool kit from here forward: the importance of constantly putting myself in the readers' shoes, making sure I'm not just addressing my own needs in a book, but theirs too. This sounds simple—but simple does not mean easy. Any success in this effort is in large part owed to her. All the failures are mine. Gideon Weil seamlessly took the baton for a period of time when both Anna and I had new arrivals to our respective families. Chantal Tom helped at every step of the way to manage this entire project and ensure all our i's were dotted and t's were crossed. Aly Mostel, Ann Edwards, and Louise Braverman for the wonderful marketing and publicity (unfortunately books do not sell themselves; fortunately I had such an amazing team to help). The unsung heroes behind every book made mine much better too: copyeditor Tanya Fox, production editor Mary Grangeia, and cover designer Stephen Brayda.

To everyone whose story appears in this book, thank you! To everyone whose research appears in this book, thank you! To my coaching clients, for ensuring everything I write about genuinely works when the rubber meets the road, thank you! And to all my readers, thank you! We are all just figuring it out, as best as we can, as we go. I am fortunate and honored to have you with me, walking the path together.

Appendix: List of Common Core Values

- Achievement
- Adventure
- Appreciation
- Attentiveness
- Authenticity
- Authority
- Autonomy
- Balance
- Beauty
- Belonging
- Boldness
- Building
- Challenge
- Citizenship
- Community
- Compassion
- Competency
- Consistency
- Contribution
- Craft
- Creativity
- Curiosity
- Determination
- Diligence
- Discernment
- Discipline
- Drive
- Effectiveness
- Efficiency
- Empathy
- Fairness
- Friendship
- Fun
- Growth
- Happiness
- Honesty
- Humility
- Humor
- Intellect
- Justice
- Kindness
- Knowledge
- Leadership
- Learning
- Love
- Loyalty
- Mastery
- Meaning
- Openness
- Optimism
- Patience
- Performance
- Persistence
- Poise
- Practice
- Quality
- Recognition
- Reputation
- Respect
- Responsibility
- Security
- Service
- Skillfulness
- Stability
- Status
- Success
- Sustainability
- Temperance
- Trust
- Wealth
- Wisdom

Additional Suggested Reading

Here are other books that complement and support *Master of Change*. Though it is an imperfect system, I've done my best to categorize each based on which part of *Master of Change* it aligns with most. Many of these books were mentioned in the text, and even those that weren't influenced the ideas in this book.

If this is your entry point to my work, I highly suggest reading my prior book, *The Practice of Groundedness*. It complements this book in many ways. If *Master of Change* is about how to navigate life's unfolding path, then *The Practice of Groundedness* is about how to build a solid foundation for sustainable excellence upon which all the navigating takes place.

Rugged and Flexible Mindset
- *What Is Health?* by Peter Sterling
- *How to Change* by Katy Milkman
- *Falling Upward* by Richard Rohr
- *The Structure of Scientific Revolutions* by Thomas S. Kuhn
- *Full Catastrophe Living* by Jon Kabat-Zinn
- *Radical Acceptance* by Tara Brach
- *Tao Te Ching* by Lao Tzu (Stephen Mitchell translation)
- *Epictetus: Discourses and Selected Writings* (Penguin Classics version)
- *Meditations* by Marcus Aurelius

- *The Pali Canon: In the Buddha's Words* by Bhikkhu Bodhi
- *Death* by Todd May
- *Almost Everything* by Anne Lamott
- *Antifragile* by Nassim Nicholas Taleb
- *A Guide to the Good Life* by William B. Irvine
- *What Matters Most* by James Hollis
- *The Hero with a Thousand Faces* by Joseph Campbell
- *Lost & Found* by Kathryn Schulz
- *Man's Search for Meaning* by Viktor Frankl
- *Grit* by Angela Duckworth

Rugged and Flexible Identity
- *To Have or To Be?* by Erich Fromm
- *Devotions* by Mary Oliver
- *The Art of Living* by Thich Nhat Hanh
- *Going to Pieces Without Falling Apart* by Mark Epstein
- *The Trauma of Everyday Life* by Mark Epstein
- *The Cancer Journals* by Audre Lorde
- *The Wisdom of Insecurity* by Alan Watts
- *A Liberated Mind* by Steven C. Hayes
- *The Extended Mind* by Annie Murphy Paul
- *Clash!* by Hazel Rose Markus and Alana Conner
- *Range* by David Epstein

Rugged and Flexible Actions
- *Dopamine Nation* by Anna Lembke
- *A Significant Life* by Todd May
- *Dancing with Life* by Phillip Moffitt
- *The Shallows* by Nicholas Carr
- *Do Hard Things* by Steve Magness
- *Surviving Survival* by Laurence Gonzales
- *The Hidden Spring* by Mark Solms
- *Subtract* by Leidy Klotz

- *Stumbling on Happiness* by Daniel Gilbert
- *No Cure for Being Human* by Kate Bowler
- *Unwinding Anxiety* by Judson Brewer
- *The Way of Aikido* by George Leonard
- *Life Is Hard* by Kieran Setiya

Notes

Introduction: Rugged Flexibility—A New Model for Working with Change and Thinking About Identity over Time

2   *Research shows that, on average*: Bruce Feiler, *Life Is in the Transitions: Mastering Change at Any Age* (New York: Penguin, 2020), 16.

5   *"The fixity of the internal environment"*: Frederic L. Holmes, "Claude Bernard, the 'Milieu Intérieur,' and Regulatory Physiology," *History and Philosophy of the Life Sciences* 8, no. 1 (1986): 3–25, https://www.jstor.org/stable/23328847?seq=1.

10 *"The key goal of regulation"*: Peter Sterling, *What Is Health?: Allostasis and the Evolution of Human Design* (Cambridge, MA: MIT, 2020), xi.

10 *Sterling and Eyer first described*: Sung W. Lee, "A Copernican Approach to Brain Advancement: The Paradigm of Allostatic Orchestration," *Frontiers in Human Neuroscience* 13 (2019), https://pubmed.ncbi.nlm.nih.gov/31105539.

10 *One of the founders of modern psychology*: David J. Leigh, "Carl Jung's Archetypal Psychology, Literature, and Ultimate Meaning," *Ultimate Reality and Meaning* 34, no. 1–2 (March 2011): 95–112, https://utpjournals.press/doi/abs/10.3138/uram.34.1-2.95.

11 *The unfreezing period is often chaotic*: Syed Talib Hussain et al., "Kurt Lewin's Change Model: A Critical Review of the Role of Leadership and Employee Involvement in Organizational Change," *Journal of Innovation & Knowledge* 3, no. 3 (September–December 2018): 123–27, https://www.sciencedirect.com/science/article/pii/S2444569X16300087.

17 *Overwhelming science demonstrates:* Bruce S. McEwen, "Allostasis and Allostatic Load: Implications for Neuropsychopharmacology," *Neuropsychopharmacology* 22 (2000): 108–24, https://www.nature.com/articles/1395453.

Chapter 1: Open to the Flow of Life

23 *"I had just killed somebody"*: Hayden Carpenter, "*The Dawn Wall* Is a Great, But Incomplete, Climbing Film," *Outside*, September 18, 2018.

25 *"I cannot make the suit out":* Jerome S. Bruner and Leo Postman, "On the Perception of Incongruity: A Paradigm," *Journal of Personality* 18, no. 2 (December 1949): 206–23, https://psychclassics.yorku.ca/Bruner/Cards.

25 *Perhaps more than anything, participants:* Thomas S. Kuhn, *The Structure of Scientific Revolutions* (Chicago: University of Chicago Press, 2012), 112–13.

25 *In the time since they were published, numerous others:* Barbara Wisse and Ed Sleebos, "When Change Causes Stress: Effects on Self-Construal and Change Consequences," *Journal of Business and Psychology* 31, vol. 2 (June 2016): 249–64, https://link.springer.com/article/10.1007/s10869-015-9411-z.

26 *In it, Tzu depicts life:* Lao Tzu, *Tao Te Ching: A New English Version*, trans. Stephen Mitchell (New York: Harper Perennial, 2006), 16.

26 *Suffering, Epictetus taught:* Epictetus, *Epictetus: Discourses and Selected Writings*, trans. and ed. Robert Dobbin (New York: Penguin, 2008), 178–85.

27 *Science shows that when you chronically fight change:* Stephan J. Guyenet, *The Hungry Brain: Outsmarting the Instincts That Make Us Overeat* (New York: Flatiron Books, 2018), 205.

27 *Meanwhile, the same modern science:* Kelly McGonigal, *The Upside of Stress: Why Stress Is Good for You, and How to Get Good at It* (New York: Avery, 2016).

28 *Sensing his masterwork might cause a backlash:* Nicolaus Copernicus, dedication of *Revolutions of the Heavenly Bodies* to Pope Paul III, 1543, https://hti.osu.edu/sites/hti.osu.edu/files/dedication_of_the_revolutions_of_the_heavenly_bodies_to_pope_paul_iii_0.pdf.

29 *To his relief, the Church did not ban it:* Encyclopedia Britannica Online, s.v. "Nicolaus Copernicus," accessed October 12, 2022, https://www.britannica.com/biography/Nicolaus-Copernicus/Publication-of-De-revolutionibus.

29 *In that same year, the Church's prohibition:* Nicholas P. Leveillee, "Copernicus, Galileo, and the Church: Science in a Religious World," *Inquiries* 3, no. 5 (2011): 2, http://www.inquiriesjournal.com/articles/1675/2/copernicus-galileo-and-the-church-science-in-a-religious-world.

30 *Nonetheless, Kuhn observed:* Kuhn, *Structure of Scientific Revolutions*, 93–94.

33 *A wave of dizziness washed over Caldwell:* Tommy Caldwell, *The Push: A Climber's Journey of Endurance, Risk, and Going Beyond Limits to Climb the Dawn Wall* (New York: Penguin, 2017), 123.

35 *"If I am what I have":* Erich Fromm, *To Have or To Be?* (New York: Harper & Row, 1976), 109.

36   *The main argument:* Fromm, *To Have*, 109–10.
38   *Toward the end of* To Have or To Be?: Fromm, *To Have*, 119.
39   *"We are more likely to look for":* Daniel Gilbert, *Stumbling on Happiness*, (New York: Alfred A. Knopf, 2006).
42   *"This is, like, the hardest thing":* Hayden Carpenter, "What *The Dawn Wall* Left Out," *Outside*, September 18, 2018, https://www.outsideonline.com/culture/books-media/dawn-wall-documentary-tommy-caldwell-review.
42   *On Wednesday, January 14:* John Branch, "Pursuing the Impossible, and Coming Out on Top," *New York Times*, January 14, 2015, https://www.nytimes.com/2015/01/15/sports/el-capitans-dawn-wall-climbers-reach-top.html.
42   *The* New York Times *put it best:* Branch, "Pursuing the Impossible."
43   *And it is the certainty of loss:* Todd May, *Death (The Art of Living)* (London: Routledge, 2016).

Chapter 2: Expect It to Be Hard

47   *Their findings, which were published:* Kaare Christensen, Anne Maria Herskind, and James W. Vaupel, "Why Danes Are Smug: Comparative Study of Life Satisfaction in the European Union," *BMJ* 333, no. 7582 (December 2006): 1289, http://www.bmj.com/content/333/7582/1289.
48   *Whereas homeostasis is agnostic to expectations:* Peter Sterling, "Allostatis: A Model of Predictive Regulation," *Physiology & Behavior* 106, no. 1 (April 2012): 5–15, https://pubmed.ncbi.nlm.nih.gov/21684297.
49   *These predictions are sent to the brain stem:* Andy Clark, "Whatever Next?: Predictive Brains, Situated Agents, and the Future of Cognitive Science," *Behavioral and Brain Sciences* 36, no. 3 (June 2013): 181–204, doi:10.1017/S0140525X12000477.
50   *"The prediction component brings out the idea":* India Morrison, Irene Perini, and James Dunham, "Facets and Mechanisms of Adaptive Pain Behavior: Predictive Regulation and Action," *Frontiers in Human Neuroscience* 7 (October 2013), https://www.frontiersin.org/articles/10.3389/fnhum.2013.00755/full.
50   *But it was the fact that the conditions gradually got better:* Daniel Kahneman et al., "When More Pain Is Preferred to Less: Adding a Better End," *Psychological Science* 4, no. 6 (November 1993): 401–5, https://www.jstor.org/stable/40062570.
50   *For example, most people rate more positively:* Ziv Carmon and Daniel Kahneman, "The Experienced Utility of Queuing: Experience Profiles and Retrospective Evaluations of Simulated Queues" (PhD diss., Duke University and Princeton University), https://www.researchgate.net/publication/236864505.

51 *For instance, when fatigued and crashing athletes:* Noel E. Brick et al., "Anticipated Task Difficulty Provokes Pace Conservation and Slower Running Performance," *Medicine & Science in Sports & Exercise* 51, no. 4 (April 2019): 734, https://journals.lww.com/acsm-msse/Fulltext/2019/04000/anticipated_task_difficulty_provokes_pace.16.aspx.

51 *The calories and nutrients:* Thays de Ataide e Silva et al., "Can Carbohydrate Mouth Rinse Improve Performance During Exercise?: A Systematic Review," *Nutrients* 6, no. 1 (January 2014): 1–10, https://www.ncbi.nlm.nih.gov/pmc/articles/PMC3916844.

55 *Hollerbach didn't view:* A'Dora Phillips, "There Is Such a Thing as Instinct in a Painter," The Vision & Art Project, January 27, 2017, https://visionandartproject.org/features/serge-hollerbach.

56 *Frankl is well-known for his book:* Viktor E. Frankl, *Man's Search for Meaning* (Boston: Beacon Press, 2006).

57 *In a study including over seventy-thousand individuals:* Emma L. Bradshaw et al., "A Meta-analysis of the Dark Side of the American Dream: Evidence for the Universal Wellness Costs of Prioritizing Extrinsic over Intrinsic Goals," *Journal of Personality and Social Psychology* (2022), http://psycnet.apa.org/record/2022-90266-001.

58 *This, of course, is exactly what Frankl did:* Frankl, *Man's Search for Meaning*.

59 *They experienced the same levels:* Barbara L. Fredrickson et al., "What Good Are Positive Emotions in Crises?: A Prospective Study of Resilience and Emotions Following the Terrorist Attacks on the United States on September 11th, 2001," *Journal of Personality and Social Psychology* 84, no. 2 (February 2003): 365–76, https://www.ncbi.nlm.nih.gov/pmc/articles/PMC2755263.

60 *Put them together and what you get:* John Maher, "When Siddartha Met Sigmund: *PW* Talks with Mark Epstein," *Publishers Weekly*, December 15, 2017, https://www.publishersweekly.com/pw/by-topic/authors/interviews/article/75640-when-siddartha-met-sigmund-pw-talks-with-mark-epstein.

61 *If, however, we can respond skillfully:* McGonigal, *Upside of Stress*.

62 *"I believe that we are all":* "*Just Mercy* Interview with Bryan Stevenson," Rolling Out, December 17, 2019, video, 6:57, https://www.youtube.com/watch?v=vZZ6xp38ukM.

63 *Innovation, creativity, and development:* "Bryan Stevenson: We Need to Talk about an Injustice," TED, March 5, 2012, video, 23:41, https://www.youtube.com/watch?v=c2tOp7OxyQ8; and "*Just Mercy* Interview."

64 *Hope, writes the moral philosopher Kieran Setiya:* Kieran Setiya, *Life Is Hard: How Philosophy Can Help Us Find Our Way* (New York: Riverhead Books, 2022), 178.

65 *Although this equation may not be mathematically perfect:* Phillip Moffitt, *Dancing with Life* (Emmaus, PA: Rodale, 2008). I first came upon the equation "suffering equals pain times resistance" in Phillip Moffitt's book. When I tried to find an original source, however, I came across multiple. The best I can offer is that this equation traces itself back to the Western Buddhist community.

67 *This graded exercise approach:* Joel Streed, "Pain Rehabilitation Center Offers Freedom from Debilitating Symptoms," Mayo Clinic, March 11, 2020, https://sharing.mayoclinic.org/2020/03/11/pain-rehabilitation-center-offers-freedom-from-debilitating-symptoms.

67 *Though Jasper still experiences pain:* Support for this reporting as well as some of the direct quotes used are here. However, I've changed the identity of the person in the running the text.

Chapter 3: Cultivate a Fluid Sense of Self

76 *This is especially true:* Robert J. Vallerand et al., "*Les Passions de l'Âme*: On Obsessive and Harmonious Passion," *Journal of Personality and Social Psychology* 85 (2003): 756–67, https://selfdeterminationtheory.org/SDT/documents/2003_VallerancBlanchardMageauKoesnterRatelleLeonardGagneMacolais_JPSP.pdf.

76 *If your identity becomes too enmeshed:* Nils van der Poel, "How to Skate a 10K . . . and Also Half a 10K," accessed October 12, 2022, https://www.howtoskate.se/_files/ugd/e11bfe_b783631375f543248e271f440bcd45c5.pdf.

80 *"Growing up in the South,":* Fortune Feimster, *Sweet & Salty,* 2020, Netflix special, 1:01:00.

80 *"We are going to Hooters":* "Kelly Clarkson Cry-Laughs Hearing Fortune Feimster's Hilarious Coming Out Story," The Kelly Clarkson Show, April 3, 2020, video, 9:00, https://www.youtube.com/watch?v=Ub7_k-J-4FE&feature=youtu.be.

80 *Reflecting on that moment in a podcast interview:* Fortune Feimster, "Ginger's Thoughts on Fortune's Marriage," November 11, 2020, *Sincerely Fortune*, podcast, episode 91, 39:24, https://sincerelyfortune.libsyn.com/episode-91.

81 *"It was a challenge for me to discover":* Van der Poel, "How to Skate."

82 *"I just hope human rights":* Chris Buckley, Tariq Panja, and Andrew Das, "Swedish Olympic Star Gives Away Gold Medal to Protest Beijing's Abuses," *New York Times*, February 25, 2022, https://www.nytimes.com/2022/02/25/world/asia/nils-van-der-poel-olympic-protest.html.

83 *In short, field theory says:* Kurt Lewin, *Field Theory in Social Science: Selected Theoretical Papers* (New York: Harper, 1951).

84 *"Independent selves view themselves":* Hazel Rose Markus and Alana Conner, *Clash!: How to Thrive in a Multicultural World* (New York: Plume, 2014), xii.

84 *Put differently, Western participants initially see:* Mutsumi Imai and Dedre Gentner, "A Cross-Linguistic Study of Early Word Meaning: Universal Ontology and Linguistic Influence," *Cognition* 62, no. 2 (February 1997): 169–200, https://www.sciencedirect.com/science/article/abs/pii/S0010027796007846.

84 *"Across [our] many studies":* Markus and Conner, *Clash!*, xiii.

86 *In a crucial paper:* Duarte Araujo and Keith Davids, "What Exactly Is Acquired During Skill Acquisition?," *Journal of Consciousness Studies* 18, nos. 3–4 (January 2011): 7–23, https://www.researchgate.net/publication/233604872_what_exactly_is_acquired_during_skill_acquisition.

92 *It cherishes its own idiosyncrasies:* Le Xuan Hy and Jane Loevinger, *Measuring Ego Development* (New York: Psychology Press, 2014).

92 *It is able to integrate:* Susanne R. Cook-Greuter, "Mature Ego Development: A Gateway to Ego Transcendence?," *Journal of Adult Development* 7, no. 4 (October 2000): 227–40, https://link.springer.com/article/10.1023/A:1009511411421.

92 *Loevinger was meticulous:* Jane Loevinger, "Construct Validity of the Sentence Completion Test of Ego Development," *Applied Psychological Measurement* 3, no. 3 (July 1979): 281–311, https://conservancy.umn.edu/bitstream/handle/11299/99630/1/v03n3p281.pdf; and Shash Ravinder, "Loevinger's Sentence Completion Test of Ego Development: A Useful Tool for Cross-Cultural Researchers," *International Journal of Psychology* 21, no. 1–4 (February–December 1986): 679–84, https://www.tandfonline.com/doi/abs/10.1080/00207598608247614?journalCode=pijp20.

94 *The Buddha's silence means:* "Ananda, Is There a Self?," in *Connected Discourses on the Undeclared*, 44.10, *Samyutta Nikaya*, Pali Canon.

95 *The ultimate self:* "Ananda Sutta: To Ananda (On Self, No Self, and Not-Self)," 2004, https://www.accesstoinsight.org/tipitaka/sn/sn44/sn44.010.than.html.

96 *"[You realize] you are not":* Nathaniel Lee, Jacqui Frank, and Lamar Salter, "Terry Crews: Here's How My NFL Career Helped and Hurt Me," *Insider*, March 22, 2017, https://www.businessinsider.com/terry-crews-heres-how-my-nfl-career-2017-3.

97 *"It's kind of weird because":* Jason Guerrasio, "How Terry Crews Went from Sweeping Floors after Quitting the NFL to Becoming a Transcendent Pitchman and Huge TV Star," *Insider*, January 18, 2018, https://www.businessinsider.com/terry-crews-sweeping-floors-to-huge-star-silence-breaker-2018-1.

97   *"I have to say, being in the NFL":* "Terry Crews Breaks Down His Career, from *White Chicks* to *Brooklyn Nine-Nine*," *Vanity Fair*, February 6, 2020, video, 21:55, https://www.vanityfair.com/video/watch/careert-timeline-terry-crews-breaks-down-his-career-from-white-chicks-to-brooklyn-nine-nine.

98   *Since* Range *was published in 2019, additional studies:* https://www.nature.com/articles/s41467-021-25477-8.

Chapter 4: Develop Rugged and Flexible Boundaries

101   *If Durante embodied any one attribute:* Danny Hajek, "Mafia Wife, Getaway Driver, Stuntwoman: From the Underworld to Hollywood," "All Things Considered," *NPR*, September 21, 2014, https://www.npr.org/2014/09/21/350120159.

102   *"Life is what it is":* Georgia Durante, *The Company She Keeps: The Dangerous Life of a Model Turned Mafia Wife* (New York: Berkley, 2008), 457–58.

105   *The individuals who reflected on their core values:* Emily B. Falk et al., "Self-Affirmation Alters the Brain's Response to Health Messages and Subsequent Behavior Change," *Proceedings of the National Academy of Sciences* 112, no. 7 (February 2015): 1977–82, https://www.pnas.org/doi/10.1073/pnas.1500247112.

106   *Hayes and his colleagues have demonstrated:* Steven C. Hayes et al., "Acceptance and Commitment Therapy and Contextual Behavioral Science: Examining the Progress of a Distinctive Model of Behavioral and Cognitive Therapy," *Behavior Therapy* 44, no. 2 (June 2013): 180–98, https://www.ncbi.nlm.nih.gov/pmc/articles/PMC3635495.

108   *"You can be stubborn and successful":* Luigi Gatto, "Roger Federer: 'You Need to Be Stubborn, Believe in Hard Work,'" *Tennis World*, August 30, 2018, https://www.tennisworldusa.org/tennis/news/Roger_Federer/59546/roger-federer-you-need-to-be-stubborn-believe-in-hard-work.

110   *In his book* The First Three Minutes*:* Steven Weinberg, *The First Three Minutes: A Modern View of the Origin of the Universe* (New York: Basic Books, 1993), 4.

110   *Initially, Wilson wasn't thrilled:* "Penzias and Wilson Discover Cosmic Microwave Radiation," *PBS*, 1965, https://www.pbs.org/wgbh/aso/databank/entries/dp65co.html.

110   *After all, he subscribed to its rival:* "Discovering the Cosmic Microwave Background with Robert Wilson," CfAPress, February 28, 2014, video, 21:55, https://youtu.be/ATaCs6Anx0c.

111   *Today, astronomers use the CMB:* "Cosmic Microwave Background," Center for Astrophysics, Harvard & Smithsonian, accessed October 12,

2022, https://pweb.cfa.harvard.edu/research/topic/cosmic-microwave-background.

111 *The duo's rugged flexibility was rewarded:* "The Nobel Prize in Physics 1978," Nobel Prize Organisation, https://www.nobelprize.org/prizes/physics/1978/summary.

112 *Third, the bigger the external change:* Michael T. Hannan and John Freeman, "The Population Ecology of Organizations," *American Journal of Sociology* 82, no. 5 (March 1977): 929–64, https://www.jstor.org/stable/2777807.

114 *And since 2004, employment in the newspaper sector:* "Newspaper Fact Sheet," Pew Research Center, June 29, 2021, https://www.pewresearch.org/journalism/fact-sheet/newspapers.

114 *In the year 2000:* Amy Watson, "Average Paid and Verified Weekday Circulation of *The New York Times* from 2000 to 2021, (in 1,000 copies)" Statista, June 21, 2022, https://www.statista.com/statistics/273503/average-paid-weekday-circulation-of-the-new-york-times.

114 *By 2022, the* Times *had:* Alexandra Bruell, "*New York Times* Tops 10 Million Subscriptions as Profit Soars," *Wall Street Journal*, February 2, 2022, https://www.wsj.com/articles/new-york-times-tops-10-million-subscriptions-as-profit-soars-11643816086.

114 *In 2021, the* Times *reported:* "The New York Times Company 2021 Annual Report," March 11, 2022, https://nytco-assets.nytimes.com/2022/03/The-New-York-Times-Company-2021-Annual-Report.pdf.

115 *As early as 1994, the* Times*'s publisher:* Gabriel Snyder, "*The New York Times* Claws Its Way into the Future," *WIRED*, February 12, 2017, https://www.wired.com/2017/02/new-york-times-digital-journalism.

115 *Here is the* Times *executive editor:* Dean Baquet, "#398: Dean Baquet," June 2020, *Longform*, podcast, episode 398, 1:34:31, https://longform.org/posts/longform-podcast-398-dean-baquet.

116 *To fully understand how science progresses through uncertainty:* Kuhn, *Structure of Scientific Revolutions*, 184, 198.

117 *As such, Lorde became an essential figure:* "Audre Lorde," Poetry Foundation, accessed October 12, 2022, https://www.poetryfoundation.org/poets/audre-lorde.

118 *"I carry death around in my body":* Audre Lorde, *The Cancer Journals*, (New York: Penguin, 2020), 5–30.

119 *He called this our* continuation body: Thich Nhat Hanh, *The Art of Living: Peace and Freedom in the Here and Now* (San Francisco: HarperOne, 2017), 71.

119 *Our actions represent:* Thich Nhat Hanh, *Understanding Our Mind: 50 Verses on Buddhist Psychology* (Berkeley, CA: Parallax Press, 2002).

119　*This means that, particularly during periods of disorder:* William MacAskill, *What We Owe the Future* (New York: Basic Books, 2022), 43.

Chapter 5: Respond Not React

125　*Over two thousand years ago:* Epictetus, *A Selection from the Discourses of Epictetus with the Encheiridion*, trans. George Long, January 9, 2004, http://pioneer.chula.ac.th/~pukrit/bba/Epictetus.pdf.

125　*In the foundational Taoist text* Tao Te Ching*:* Mitchell, *Tao Te Ching*, 45.

131　*"Zanshin is the future":* George Leonard, *The Way of Aikido: Life Lessons from an American Sensei* (New York: Plume, 2000), 120–23.

131　*"If an unexpected object enters your vision":* "Target Fixation: It's Not Just a Motorcycle Problem," Drive Safely, accessed October 12, 2022, https://www.idrivesafely.com/defensive-driving/trending/target-fixation-its-not-just-motorcycle-problem.

133　*Park attributes her proficiency:* Max Schreiber, "Inbee Park Explains Why She's the World's Best Putter from 10–15 Feet," GOLF Channel, October 6, 2021, https://www.golfchannel.com/news/inbee-park-explains-why-shes-worlds-best-putter-1015-feet.

133　*"I came to realize that":* Brentley Romine, "This Mental Tip from LPGA Legend Inbee Park Is Major," GOLF Channel, January 21, 2022, https://www.golfchannel.com/news/mental-tip-lpga-legend-inbee-park-major.

134　*You can think of the striatum:* José L. Lanciego, Natasha Luquin, and José A. Obeso, "Functional Neuroanatomy of the Basal Ganglia," *Cold Spring Harbor Perspectives in Medicine* 2, no. 12 (December 2012): a009621, https://www.ncbi.nlm.nih.gov/pmc/articles/PMC3543080.

135　*His work (and that of other investigators) shows:* Kenneth L. Davis and Christian Montag, "Selected Principles of Pankseppian Affective Neuroscience," *Frontiers in Neuroscience* 12 (January 2019), https://www.frontiersin.org/articles/10.3389/fnins.2018.01025/full.

135　*If you are able to muster a deliberate response:* Guyenet, *Hungry Brain*, 205.

135　*It makes us feel good:* Andrew B. Barron, Eirik Søvik, and Jennifer L. Cornish, "The Roles of Dopamine and Related Compounds in Reward-Seeking Behavior across Animal Phyla," *Frontiers in Behavioral Neuroscience* 4 (October 2010), https://www.frontiersin.org/articles/10.3389/fnbeh.2010.00163/full.

136　*The SEEKING pathway, and the dopamine that fuels it:* Mark Solms, *The Hidden Spring* (New York: W. W. Norton, 2021), 89.

136　*Meanwhile, if a longer period:* Craig N. Sawchuk, "Depression and Anxiety: Can I Have Both?," Mayo Clinic, June 2, 2017, https://www

.mayoclinic.org/diseases-conditions/depression/expert-answers/depression-and-anxiety/faq-20057989.

137 *This is why depression is characterized:* Solms, *Hidden Spring*, 115.

137 *I suspect this is also why behavioral therapies:* Sona Dimidjian et al., "The Origins and Current Status of Behavioral Activation Treatments for Depression," *Annual Review of Clinical Psychology* 7 (2011): 1–38, https://pubmed.ncbi.nlm.nih.gov/21275642.

139 *"When Karla was barely thirteen":* "Cristina Martinez," *Chef's Table*, volume 5, episode 1, September 28, 2018, Netflix special, 50:00.

140 *And therein lay her calculated response:* "Barbacoa sin fronteras (Barbacoa Beyond Borders)," December 2, 2021, *Duolingo*, podcast, episode 100, 23:46, https://podcast.duolingo.com/episode-100-barbacoa-sin-fronteras-barbacoa-beyond-borders.

141 *"All of a sudden, I'm on the radio":* "Cristina Martinez."

142 *Decades of research show:* Albert Bandura, "Self-Efficacy: Toward a Unifying Theory of Behavioral Change," *Psychological Review* 84, no. 2 (March 1977): 191–215, https://psycnet.apa.org/doiLanding?doi=10.1037%2F0033-295X.84.2.191.

144 *In other words, it is the act of labeling:* Matthew D. Lieberman et al., "Putting Feelings into Words: Affect Labeling Disrupts Amygdala Activity in Response to Affective Stimuli," *Psychological Science* 18, no. 5 (May 2007): 421–28, https://pubmed.ncbi.nlm.nih.gov/17576282.

145 *For example, in Scandinavian folklore:* Francis James Child, *The English and Scottish Popular Ballads*, vol. 1 (New York: Dover Publications, 1965), 95–96.

145 *The only way the saint could free himself:* Child, *Popular Ballads*, 95.

145 *In arguably the most widely known example:* Maria Tatar, ed., *The Annotated Classic Fairy Tales* (New York: W. W. Norton, 2002), 128.

146 *Non-identify with your experience:* Tara Brach, "Feeling Overwhelmed? Remember RAIN," *Mindful*, February 7, 2019, https://www.mindful.org/tara-brach-rain-mindfulness-practice.

146 *Research shows that this is true:* D. M. Perlman et al., "Differential Effects on Pain Intensity and Unpleasantness of Two Meditation Practices," *Emotion* 10, no. 1 (February 2010): 65–71, https://doi.org/10.1037/a0018440; the website for the UMass Memorial Health Center for Mindfulness, accessed October 12, 2022, https://www.umassmed.edu/cfm/research/publications; Philippe R. Goldin and James J. Gross, "Effects of Mindfulness-Based Stress Reduction (MBSR) on Emotion Regulation in Social Anxiety Disorder," *Emotion* 10, no. 1 (February 2010): 83–91, doi:10.1037/a0018441; and Igor Grossmann and Ethan Kross, "Exploring Solomon's Paradox: Self-Distancing Eliminates the Self-Other Asymmetry in Wise Reasoning About Close Relationships in Younger and Older

Adults," *Psychological Science* 25, no. 8 (August 2014): 1571–80, doi:10.1177/0956797614535400.

146 *When you find yourself confronted:* Özlem Ayduk and Ethan Kross, "From a Distance: Implications of Spontaneous Self-Distancing for Adaptive Self-Reflection," *Journal of Personality and Social Psychology* 98, no. 5 (May 2010): 809–29, https://www.ncbi.nlm.nih.gov/pmc/articles/PMC2881638.

147 *If reacting and the associated RAGE pathway:* Aldous Huxley's "cerebral reducing value": https://www.researchgate.net/figure/Aldous-Huxleys-cerebral-reducing-valve-on-the-inlet-right-side-of-the-cerebral_fig1_323345114.

148 *Dacher Keltner, a professor of psychology:* Jennifer E. Stellar et al., "Awe and Humility," *Journal of Personality and Social Psychology* 114, no. 2 (February 2018): 258–69, https://sites.lsa.umich.edu/whirl/wp-content/uploads/sites/792/2020/08/2018-Awe-and-Humility.pdf.

148 *According to a 2015 study:* Jennifer E. Stellar et al., "Positive Affect and Markers of Inflammation: Discrete Positive Emotions Predict Lower Levels of Inflammatory Cytokines," *Emotion* 15, no. 2 (April 2015): 129–33, https://www.ncbi.nlm.nih.gov/pubmed/25603133.

148 *"Adults spend more time working and commuting":* Dacher Keltner, "Why Do We Feel Awe?," *Mind & Body*, May 10, 2016, https://greatergood.berkeley.edu/article/item/why_do_we_feel_awe.

148 *But when they are given a compound:* Yukiori Goto and Anthony A. Grace, "Dopaminergic Modulation of Limbic and Cortical Drive of Nucleus Accumbens in Goal-Directed Behavior," *Nature Neuroscience* 8, no. 6 (June 2005): 805–12, http://www.nature.com/neuro/journal/v8/n6/full/nn1471.html.

149 *Research in continuous improvement:* "What Is the Plan-Do-Check-Act (PDCA) Cycle?," ASQ, accessed October 12, 2022, https://asq.org/quality-resources/pdca-cycle.

149 *In 1964, the Canadian communication theorist Marshall McLuhan:* Marshall McLuhan, *Understanding Media: The Extensions of Man* (CreateSpace, 2016), https://web.mit.edu/allanmc/www/mcluhan.mediummessage.pdf.

150 *Moreover, research shows the two factors:* Jonathan Haidt, "Why the Past 10 Years of American Life Have Been Uniquely Stupid," *The Atlantic*, April 11, 2022, https://www.theatlantic.com/magazine/archive/2022/05/social-media-democracy-trust-babel/629369.

Chapter 6: Making Meaning and Moving Forward

156 *"Such rooms in our common psychic mansion":* James Hollis, *What Matters Most: Living a More Considered Life* (New York: Avery, 2009), 147.

Notes 217

159 *In these circumstances, premature attempts:* Daniel Gilbert, *Stumbling Upon Happiness: Think You Know What Makes You Happy?* (New York: Knopf, 2006), 191.

160 *As such, they don't feel as devastatingly long:* Adrian Bejan, "Why the Days Seem Shorter as We Get Older," *European Review* 27, no. 2 (May 2019): 187–94, doi:10.1017/S1062798718000741.

160 *This explains why many evidence-based therapies:* "Post-Traumatic Stress Disorder (PTSD)," Mayo Clinic, July 16, 2018, https://www.mayoclinic.org/diseases-conditions/post-traumatic-stress-disorder/symptoms-causes/syc-20355967.

161 *But when Eagleman asked participants:* Chess Stetson, Matthew P. Fiesta, and David M. Eagleman, "Does Time Really Slow Down During a Frightening Event?," *PLoS One* 2, no. 12 (2007): e1295, https://journals.plos.org/plosone/article?id=10.1371/journal.pone.0001295.

162 *The "recency bias" says that:* Gilbert, *Stumbling Upon Happiness*, 170–73.

162 *In a series of studies:* Daniel T. Gilbert, Erin Driver-Linn, and Timothy D. Wilson, "The Trouble with Vronsky: Impact Bias in the Forecasting of Future Affective States," in *The Wisdom in Feeling: Psychological Processes in Emotional Intelligence*, eds. Lisa Feldman Barrett and Peter Salovey (New York: Guilford Press, 2002).

163 *The researchers go on to write:* Timothy D. Wilson and Daniel T. Gilbert, "Affective Forecasting: Knowing What to Want," *Current Directions in Psychological Science* 14, no. 3 (June 2005): 131–34, https://www.jstor.org/stable/20183006.

167 *Research shows that the most common outcome:* George A. Bonanno, Courtney Rennicke, and Sharon Dekel, "Self-Enhancement Among High-Exposure Survivors of the September 11th Terrorist Attack: Resilience or Social Maladjustment?," *Journal of Personality and Social Psychology* 88, no. 6 (June 2005): 984–98, https://pubmed.ncbi.nlm.nih.gov/15982117.

167 *Interestingly, but not at all surprising:* Terri A. deRoon-Cassini et al., "Psychopathy and Resilience Following Traumatic Injury: A Latent Growth Mixture Model Analysis," *Rehabilitation Psychology* 55, no. 1 (February 2010): 1–11, doi:10.1037/a0018601.

168 *And yet, when you look across the literature:* Richard G. Tedeschi and Lawrence G. Calhoun, "Posttraumatic Growth: Conceptual Foundations and Empirical Evidence," *Psychological Inquiry* 15, no. 1 (2004): 1–18, doi:10.1207/s15327965pli1501_01.

169 *When someone feels lost or broken:* "Anna Lembke on the Neuroscience of Addiction: Our Dopamine Nation," Rich Roll, August 23,

2012, video, 2:18:02, minutes 48–52, https://www.youtube.com/watch?v=jziP0CegvOw.

170 *"If we try to control a situation or our lives"*: Judson Brewer, *The Craving Mind: From Cigarettes to Smartphones to Love—Why We Get Hooked and How We Can Break Bad Habits* (New Haven, CT: Yale University Press, 2017), 111.

171 *Studies show that asking for:* Melanie A. Hom et al., "Resilience and Attitudes Toward Help-Seeking as Correlates of Psychological Well-Being Among a Sample of New Zealand Defence Force Personnel," *Military Psychology* 32, no. 4 (2020): 329–40, https://www.tandfonline.com/doi/abs/10.1080/08995605.2020.1754148; and Allison Crowe, Paige Averett, and J. Scott Glass, "Mental Illness Stigma, Psychological Resilience, and Help Seeking: What Are the Relationships?," *Mental Health & Prevention* 4, no. 2 (June 2016): 63–68, https://www.sciencedirect.com/science/article/abs/pii/S2212657015300222.

173 *Instead, participants immediately assume:* Gabrielle S. Adams et al., "People Systematically Overlook Subtractive Changes," *Nature* 592, no. 7853 (April 2021): 258–61, https://www.nature.com/articles/s41586-021-03380-y.

174 *Research shows that the same brain region:* Nora D. Volkow et al., "Cocaine Cues and Dopamine in Dorsal Striatum: Mechanism of Craving in Cocaine Addiction," *Journal of Neuroscience* 26, no. 24 (June 2006): 6583–88, https://www.jneurosci.org/content/26/24/6583.

174 *"[Rituals] open a space in which to host thoughts"*: Katherine May, *Wintering: The Power of Rest and Retreat in Difficult Times* (New York: Riverhead Books, 2020), 115.

175 *In his book outlining allostasis:* Sterling, *What Is Health?*, 102.

179 *"Not wore out. Wore* down*"*: Robert M. Pirsig, *Zen and the Art of Motorcycle Maintenance: An Inquiry into Values* (New York: William Morrow, 1974), 243.

181 *But we can transform suffering:* Thich Nhat Hanh, *No Mud, No Lotus: The Art of Transforming Suffering* (Berkeley, CA: Parallax Press, 2014).

181 *In a multiyear study of more than 2,000 adults:* Mark D. Seery, Alison Holman, and Roxane Cohen Silver, "Whatever Does Not Kill Us: Cumulative Lifetime Adversity, Vulnerability, and Resilience," *Journal of Personality of Social Psychology* 99, no. 6 (December 2010): 1025–41, doi: 10.1037/a0021344.

182 *We shake with joy, we shake with grief:* Mary Oliver, *Devotions: The Selected Poems of Mary Oliver* (New York: Penguin, 2017), 70.

182 *"There is enlargement of the spirit"*: Hollis, *What Matters Most*, 163.

Conclusion: Five Questions and Ten Tools for Embracing Change and Developing Rugged Flexibility

185 *In the middle of 2022:* Rebecca Solnit, "Why Did We Stop Believing That People Can Change?," *New York Times*, April 22, 2022, https://www.nytimes.com/2022/04/22/opinion/forgiveness-redemption.html.

188 *There is a story of a wise Thai Forest elder:* Mark Epstein, *Thoughts Without a Thinker: Psychotherapy from a Buddhist Perspective* (New York: Basic Books, 2013), 79–81.

191 *As the Nobel Prize–winning psychologist Danny Kahneman:* Matthew Hutson, "Why Our Efficient Minds Make So Many Bad Errors," *Washington Post*, December 9, 2016, https://www.washingtonpost.com/opinions/why-our-efficient-minds-make-so-many-bad-errors/2016/12/08/4eb98fce-b439-11e6-840f-e3ebab6bcdd3_story.html.

193 *In an interview with* The Atlantic: David Brooks, "Bruce Springsteen and the Art of Aging Well," *The Atlantic*, October 23, 2020, https://www.theatlantic.com/ideas/archive/2020/10/bruce-springsteen-and-art-aging-well/616826.

193 *As the Harvard psychologist turned spiritual teacher:* Ram Dass and Mirabai Bush, *Walking Each Other Home: Conversations of Loving and Dying* (Boulder, CO: Sounds True, 2018).

Index

acceptance and commitment
    therapy (ACT), 10–11, 105
actions. *See also* meaning and
    growth; responding to
    change
  behavioral activation and, 138,
    196–97
  as belongings, 119
  brain and, 174
  as experiments, 149
  letting go, 168–70
  routines and, 174
  SEEKING pathway and, 137–38
  wise, 60–61, 68, 193–94
activation energy, 138
adversity. *See also* suffering
  anxiety and, 156
  as inevitable, 152
  meaning in (*See* meaning and
    growth)
  outcomes for levels of, 181–82
  prediction and, 163–64
  time distorted during, 159–61
affect, 136, 137–38
affect labeling, 144–45, 195
agency, 7, 64. *See also* actions
allostasis, 8–9, 10–11, 12, 48
amygdala, 134
Ananda, 93, 94
anatta, 40
anger, 135, 136–37
anicca, 40

anxiety, 11, 105–6, 156
Araujo, Duarte, 86
Ashman, Jay, 163–67, 171
awe, 147–48, 175

Baquet, Dean, 115
basal ganglia, 134–35
behavioral activation, 138, 176, 177,
    178, 196–97
behavioral therapies, 137
being orientation, 36, 192
Bernard, Claude, 5
bias, 161–62, 163
Bodhi, Bhikkhu, 93, 94
boundaries, 120–21
  core values and, 103–7, 120
  flexible application of, 107–9
  flexible sense of self and, 100, 104
  for organizations, 112–13
Boyd, Sophia Alvarez, 86
brain function
  boundaries and, 104–5
  expecting challenges, 49–53,
    66–67
  making meaning and, 160–61,
    174, 175
  neuroplasticity, 9
  responding vs. reacting, 134–38,
    148–49
Brewer, Judson, 170
Brown, David, 66–67
Bruner, Jerome, 24–25, 27

Buddha, 93–94
Buddhism, 26, 27, 59–60, 93–94, 95
burnout, 176

cable news, 150–51
Caldwell, Tommy, 21–24, 32–35, 36, 38–39, 41–42
Cancer Journals, The (Lorde), 118
Cannon, Walter, 5
"Case for Tragic Optimism" (Frankl), 56
Chaa, Achaan, 188
change
 ancient views of, 4, 26, 27, 125–26
 compassion and, 179–80
 cycle of (*See* cycle of order, disorder, and reorder)
 difficult (*See* adversity; trauma)
 as disruptive, 24, 30–32
 fear of, 31
 growth from (*See* meaning and growth)
 inevitability of, 3, 14, 40–41, 74, 82–83, 185, 186
 loss and, 40–41, 43
 as neutral, 4
 non-dual thinking and, 13–14, 185
 pace of, 113
 political movements and, 31–32
 time distortion and, 159–63
 truth and, 74
 types of, 40
 Western views of, 5–6, 7
change, responses to
 accepting, 24–26, 27–28, 32, 39, 185, 186–90
 adapting to, 182, 185, 186–90
 assessing control, 27–28, 60–61
 avoiding and denying, 6, 60
 reacting (*See* reacting to change)
 resisting (*See* resisting change)
 responding (*See* responding to change)
 sacrificing agency, 7
 striving for past, 7, 30–31, 32
Christine (client), 36–39
Clark, Andy, 49
Clash! (Markus, Connor), 84–85
community, 181, 193
Company She Keeps, The (Durante), 102
compassion, 147, 179–81, 193
complexity, 78, 82, 194
Conner, Alana, 83–85
consciousness, 49, 51, 52
control
 lack of, 125–26, 168
 letting go of, 168–70
 of response, 58, 125–26
 of situation, 26, 27–28, 60–61
conventional self, 94–95
Copernicus, Nicolaus, 28–29
core values
 boundaries and, 103–9, 117, 189–90
 defined, 103
 identifying, 106, 189
 responding and, 138, 143
 shifting of, 188
cortisol, 27, 61
COVID-19 pandemic, 2, 30, 31, 45–46, 53, 126–30
Crews, Terry, 96–97
Cruz, Nicola, 85–86
curiosity, 147
cycle of order, disorder, and reorder, 120. *See also* allostasis
 compassion and, 181
 in development of self, 11
 in evolution, 77–78
 meaning and, 157
 in scientific progress, 30, 113, 116

Dante Alighieri, 152
Dass, Ram, 193
Davis, Keith, 86

Index

Death (May), 42
death, identity and, 118
depression, 137, 172
Dialogue Concerning the Two Chief World Systems (Copernicus), 29
Dicke, Robert, 110
Dickey, John, 21–23
differentiation, 78, 194
disorder events, 2
dopamine, 135, 136, 137, 148–49, 174
dukkha, 59–60
Durante, Georgia, 100–103, 104

Eagleman, David, 160–61
ego, 77, 91–93, 95, 147
emotions, 41, 58–59, 144–45, 162, 172, 195. *See also* RAGE pathway; SADNESS pathway
environment, 83–85, 86, 87
Epictetus, 26, 125
Epstein, David, 97–98
Epstein, Mark, 11
Equal Justice Initiative, 61–62
Erikson, Erik, 91
evolution, 10, 77–78
expectations, 70
   accepting suffering and, 64–68
   happiness and, 46–48, 52, 56–57, 68, 187
   neuroscience of, 49–53
   realistic, 53, 59, 68, 187–88, 192
   reality shaped by, 46–48, 51–52
   tragic optimism and, 55–56, 57–59, 67–68, 193
   wise hope/wise action and, 60–61, 68, 193–94
Eyer, Joseph, 8, 9, 10

fake fatigue, 176, 177, 178
Falk, Emily, 104–5, 106
fatigue, 176–78
Federer, Roger, 107–9

Feimster, Ginger and Fortune, 78–81, 95
field theory, 83
First Three Minutes, The (Weinberg), 110
4 Ps heuristic, 143–47
Frankl, Viktor, 55–56, 58
Freeman, John, 112, 113
Freud, Sigmund, 39–40
Fromm, Erich, 35–36, 38
fundamental spiritual pivot, 169–70

Galileo Galilei, 29
generalist approach, 97–99
Gentner, Dedre, 84
Gilbert, Daniel, 39, 158, 159
gratitude, 156–57
grief, 172
growth. *See* meaning and growth
growth mindset, 156
Gui, Angela, 82

Hanh, Thich Nhat, 119, 181
Hannan, Michael, 112, 113
happiness, 46–48, 52, 56–57, 68, 187
having orientation, 36, 192
Hayes, Steven, 105, 106
heliocentrism, 29
help, asking for and receiving, 170–73
Hidden Spring, The (Solms), 135–36
Hohwy, Jakob, 49
Hollerbach, Serge, 53–55, 58
Hollis, James, 156, 168, 182
homeostasis, 5–6, 8–9, 12, 48
hope, 64, 185. *See also* wise hope and wise action
humility, 169, 170
Huxley, Aldous, 147–48

identity. *See also* boundaries; self, fluid sense of
   of organizations, 112–13
   practices supporting, 117

as stable and changing, 11
term usage, 77
transient nature of, 40
Imai, Mutsumi, 84
impact bias, 163
independence, 84–85, 87, 194–95
inescapability trigger, 39
integration, 78, 118, 188–89, 194
interdependence, 84–85, 87, 194–95

Jasper, Cathy, 66–67
joy, 41
Jung, Carl, 10, 74, 82–83
Just Mercy (Stevenson), 62

Kabat-Zinn, Jon, 173
Kahneman, Daniel, 50, 191
Katie (teacher), 126–30
Keltner, Dacher, 148
Klotz, Leidy, 173
Kuhn, Thomas, 30, 116

Lao Tzu, 26, 125, 163
Law of Affect, 136
law of names, 145
Lembke, Anna, 169
Leonard, George, 131
Lewin, Kurt, 83
Lewinsohn, Peter, 196–97
Loevinger, Jane, 91–93, 95, 147
Lorde, Audre, 117–19
loss, 40–41, 43

MacAskill, William, 119
Man's Search for Meaning (Frankl), 56
Markus, Hazel Rose, 83–85
Martinez, Cristina, 138–41, 142
May, Katherine, 175
May, Todd, 42
Mayo Clinic, 65–68
McDonald, Michele, 146

McInerny, Nora, 171–72
McLuhan, Marshall, 149–50
McMillan, Stuart, 86–87
meaning and growth, 183
allowing time for, 157–59, 162, 163, 197–98
asking for help with, 170–71, 172–73
easier with practice, 182–83
level of adversity and, 167–68, 181–82
non-duality of, 179, 185
in personal narratives, 162, 163
real *vs.* fake fatigue and, 175–78
simplicity and, 173–75
suffering and loss, 43, 181, 182
surrendering control and, 168–70
trauma, 167–68
unhelpful strategies for, 156–57
media influence, 149–51
meditation, 147
Melanie (client), 175–76
Miller, Benjamin, 140
mindset. *See also* expectations; openness to flow of life
core components of, 68–69
empowering nature of, 27, 28
external/internal change and, 69
goal of adopting, 69
motivation, 137–38, 148

newspaper industry, 113–14, 116
New York Times, the, 114–16
Niebuhr, Reinhold, 126
Nietzsche, Friedrich, 182
non-dual thinking, 13, 191

obsessive-compulsive disorder (OCD), 154–55
Oliver, Mary, 182
On the Revolutions of the Heavenly Spheres (Copernicus), 28–29

"On Transience" (Freud), 40
openness to flow of life, 44
   being orientation in, 36, 192
   change accepted in, 24–26,
     27–28, 32, 39, 186–87
   sorrow and joy in, 39–41, 43
organizational patterns, 11, 112–13

pain, 65, 66–68. *See also* suffering
panic, 136–37, 143
Panksepp, Jaak, 134, 135
Park, Inbee, 132–33
path vs. road, 16–17
pausing, 143, 144–45, 195
Penzias, Arno, 109–11
planning, 143, 145–48, 195. *See also* SEEKING pathway
population ecology, 112
posterior cingulate cortex, 170
Postman, Leo, 24–25, 27
Postman, Neil, 116
post-traumatic stress disorder (PTSD), 160, 167
Practice of Groundedness, The (Stulberg), 171
prediction, 49–53, 163–64
proceeding, 143, 148–49, 195
processing, 143, 145–48, 195
psychological fatigue, 176
psychological immune system, 158–59, 160, 162
purpose, 156–57
Push, The (Caldwell), 33–34

RAGE pathway, 134–35, 136–37, 149, 150
RAIN practice, 146
Range (Epstein), 97–98
reacting to change
   attention to conditions, 190
   as instinctive, 149, 190
   media and, 150
   neuroscience of, 134–38

real fatigue, 176–77, 178
recency bias, 161–62
resilience, 59, 167–68, 171
resisting change, 6–7
   attention to tendencies, 187
   consequences of, 4, 24, 27, 185, 187
   politics and, 31
   suffering and, 27, 60, 65, 67, 68, 187
   window of plasticity and, 119, 120
responding to change, 153
   behavioral activation, 138
   four P's for, 143–49
   media influence, 150–51
   neuroscience of, 134–38
   power of, 142–43
   time and space for, 149
   zanshin, 130–32, 133
rest, 176–77, 178
Rilke, Rainer Maria, 40
rituals, 174–75, 195–96
Rodden, Beth, 21–23, 33, 34
Rohr, Richard, 11
Rollins, Edwin, 117
routines, 173–74, 195–96
rugged flexibility
   benefits of, 3
   concept of change in, 14
   core values and, 189–90
   defined, 14
   goal of, 14, 191
   non-duality of, 13–14, 191
   tools for developing, 190–98

sacred practices, 175
SADNESS pathway, 136, 137
Schreiber, Max, 133
scientific progress, 30, 111, 116–17
SEEKING pathway, 135, 136, 137, 148, 149
Seery, Mark, 181–82

self, 77, 93–95
self, fluid sense of, 99
  vs. attachment, 76–77
  boundaries and, 100, 104
  complexity and, 77–78, 188–89, 194
  ego and, 77, 92–93, 95
  generalist approach to, 97–98
  independence and interdependence in, 84–85, 87, 194–95
  non-duality of, 85, 92, 95, 195
  stability of, 91, 93
self, sense of. *See also* ego
  differentiation and, 78, 194
  environment and, 83–85, 86
  integration and, 78, 118, 188–89, 194
  path metaphor and, 17
  response to change and, 25–26
self-compassion, 147, 179–80
self-confidence, 142, 147
self-discipline, 179. *See also* behavioral activation
self-distancing techniques, 146–48, 195
self-efficacy, 142–49
Setiya, Kieran, 64
simplicity, voluntary, 173–75
skill, 86–87, 98
Smith, Jason, 21–23
social media, 150–51
Solms, Mark, 49, 135–36, 137
Solnit, Rebecca, 185
sorrow, 41
Springsteen, Bruce, 193
Sterling, Peter, 8, 9, 10, 107, 170–71, 175, 184
Stevenson, Bryan, 61–63
striatum, 134, 174
strongman leaders, 31
Structure of Scientific Revolutions, The (Kuhn), 30, 116

suffering
  avoiding, 57–58
  compassion and, 180–81
  dukkha and, 60
  expectation-reality mismatch and, 52
  inevitability of, 58, 182
  meaning and, 40–41, 58
  vs. pain, 65
  resistance and, 27, 60, 65, 67, 68, 187
  sense of control and, 26, 60–61
  varieties of, 56
Sulzberger, Arthur Ochs, Jr., 115
summit fever, 130
surrender, 169–70
Suzuki, Shunryu, 186, 187
Sweet and Salty (2020), 80

Taoism, 26
Tao Te Ching (Lao Tzu), 26, 125, 163
target fixation, 130–31
Terrible, Thanks for Asking (podcast), 172
theory of mind, 92
time
  distortion of, 159–62
  for meaning and growth, 157–59, 197–98
To Have or To Be? (Fromm), 35–36, 38
tools, 190–98. *See also* specific tools
toxic positivity, 48, 60, 61
tragedy, varieties of, 56
tragic optimism
  defined, 57
  emotions and, 58–59
  expectations and, 59, 68
  non-duality of, 58
  origin of term, 55–56
  practicing, 193–94

stress response shaped by, 57
wise hope/wise action and, 60, 64, 193–94
trauma, 160, 167–68

ultimate self, 95
Understanding Media (McLuhan), 149–50

van der Poel, Nils, 73–76, 77, 81–83, 95
voluntary simplicity, 173–75

Weinberg, Steven, 110
What Is Health? (Sterling), 175
What We Owe the Future (MacAskill), 119
Whitman, Walt, 98
"Why Did We Stop Believing That People Can Change" (Solnit), 185
Wilson, Robert, 109–11
wisdom, 11, 193
wise hope and wise action, 60–61, 68, 193–94

zanshin, 130–32, 133
Zen and the Art of Motorcycle Maintenance (Pirsig), 178–79

About the Author

Brad Stulberg is the bestselling author of *The Practice of Groundedness* and coauthor of *Peak Performance*. Stulberg regularly contributes to *The New York Times*, and his work has been featured in *The Wall Street Journal* and *The Atlantic*, among other large outlets. He serves as the cohost of *Farewell*, a Growth Equation podcast, and is on faculty at the University of Michigan's Graduate School of Public Health. In his coaching practice, he works with executives, entrepreneurs, physicians, and athletes on their mental skills and overall well-being. He lives in Asheville, North Carolina.